Schanila Nawaz

MBP Interacts with PIP2 at the Oligodendroglial Cell Membrane

Schanila Nawaz

MBP Interacts with PIP2 at the Oligodendroglial Cell Membrane

The Role of Phosphoinositides in the Interaction of Myelin Basic Protein with the Oligodendroglial Cell Membrane

Südwestdeutscher Verlag für Hochschulschriften

Imprint
Any brand names and product names mentioned in this book are subject to trademark, brand or patent protection and are trademarks or registered trademarks of their respective holders. The use of brand names, product names, common names, trade names, product descriptions etc. even without a particular marking in this work is in no way to be construed to mean that such names may be regarded as unrestricted in respect of trademark and brand protection legislation and could thus be used by anyone.

Publisher:
Südwestdeutscher Verlag für Hochschulschriften
is a trademark of
Dodo Books Indian Ocean Ltd., member of the OmniScriptum S.R.L Publishing group
str. A.Russo 15, of. 61, Chisinau-2068, Republic of Moldova Europe
Printed at: see last page
ISBN: 978-3-8381-2079-9

Zugl. / Approved by: Göttingen, Max Planck Research School, Dissertation, 2009

Copyright © Schanila Nawaz
Copyright © 2010 Dodo Books Indian Ocean Ltd., member of the OmniScriptum S.R.L Publishing group

Acknowledgements

I would like to convey my deep gratitude to Prof. Klaus-Armin Nave, who gave me the opportunity to work in his lab and who supported me during my PhD work in every possible way. I am thankful for his supervision. I have learned a lot through him during the past three years. I also appreciate his never failing encouragement, and his valuable criticisms and feedback. I am also grateful for the use of his well-equipped laboratory.

I am sincerely grateful to PD Dr. Mikael Simons, for his relentless encouragement and advice, and most of all for the several valuable discussions I had with him. I also gratefully acknowledge the opportunity given to me to collaborate with him and the kind support I received from his group. I would like to especially thank in this connection Angelika Kippert, Larissa Yurlova and Gieselheid Schulz.

I am indebted to Prof. Reinhard Jahn and PD Dr. Evgeni Ponimaskin for their kind support and discussions during the last three years. I am also thankful to Prof. Reinhard Jahn for giving me the opportunity to work in his laboratory.

I would also like to thank Prof. Thorsten Lang for his kind support and discussions concerning membrane sheet experiments. I am grateful to Dr. Sandra Göbbels for taking a close look at my thesis.

I owe special thanks to my parents, to my sisters Anuscheh, Sehera and Ayla, and my dear friend Jan for their undaunted trust in me.

I would also like to thank Dr. Hauke Werner for taking his valuable time in discussing parts of this project.

I would also like to thank the co-ordination team from the Neuroscience Graduate Program Prof. Michael Hörner and Sandra Drube for their support in administrative matters during the last four years.

Contents

List of figures	**v**
1 Introduction	**1**
1.1 Composition and structure of myelin	1
1.2 Oligodendrocytes	2
1.3 Developmental stages of oligodendrocytes	4
1.4 Myelin basic protein	5
1.4.1 Posttranslational modifications of MBP	7
1.4.2 Myelin of shiverer (MBP-/-) mice	7
1.5 The phosphoinositide PIP2	8
1.5.1 Functions of PIP2	9
1.5.2 Enzymes generating PIP2	11
1.5.3 Protein domains binding to PIP2	12
1.5.4 Molecular tools to monitor phospholipids	12
2 Materials and Methods	**14**
2.1 Materials	14
2.1.1 Cell Culture	14
2.1.1.1 Mammalian cell lines	14
2.1.1.2 Mammalian cell culture media	14
2.1.2 Strains and cells	16
2.1.2.1 Bacterial strains	16
2.1.2.2 Bacterial culture media	16
2.1.3 Molecular cloning reagents	16
2.1.3.1 Plasmids	16
2.1.3.2 Enzymes	16
2.1.3.3 Buffers	17
2.1.3.4 Primer sequences and PCR protocol	17
2.1.4 Biochemical reagents	18
2.1.4.1 Western Blotting reagents	18
2.1.4.2 Membrane isolation buffers	19
2.1.5 Immunofluorescence labeling reagents	20

	2.1.6	cDNA Clones		20
	2.1.7	Antibodies		21
	2.1.8	Chemical compounds		21
2.2	Methods			22
	2.2.1	Manipulation of DNA		22
	2.2.2	Transformation		22
	2.2.3	Mini/Midi plasmid purification		22
	2.2.4	Generation of stable cell lines		23
	2.2.5	Biochemical techniques		23
		2.2.5.1	Sucrose gradient centrifugation	23
		2.2.5.2	Detergent resistant membrane isolation from Oli-ceu cells	23
	2.2.6	Cell culture and transfections		24
		2.2.6.1	Primary cell culture	24
		2.2.6.2	Oligodendroglial cell lines	24
	2.2.7	Expression constructs and virus generation		25
	2.2.8	Immunofluorescence staining procedure		25
	2.2.9	Life cell imaging and image analysis		26
	2.2.10	Generation of membrane sheets		26
	2.2.11	Ionomycin treatment of primary oligodendrocytes		27
	2.2.12	FRET measurement		28
	2.2.13	Quantification of protein localization at the plasma membrane		28
	2.2.14	Acute slices of corpus callosum		29
	2.2.15	Electron microscopy		30

3 Results   31

3.1 MBP accumulates at PIP2 enriched membranes 31
    3.1.1 MBP and PIP2 colocalize at the same subcellular domains 31
    3.1.2 FRET experiments indicate a close association of PIP2 with MBP . . 33
    3.1.3 PIP2 accumulation in endomembranes leads to relocalization of MBP 35
3.2 Decreased levels of PIP2 at the plasma membrane leads to decreased MBP binding 37
    3.2.1 Specific hydrolysis of PIP2 leads to reduced plasma membrane association of MBP 37
    3.2.2 PIP2 dependent plasma membrane association of MBP verified in membrane sheets 39
3.3 Decrease in PIP2 at the plasma membrane leads to intracellular accumulation of MBP 42
3.4 Replacement of positive amino acids in MBP reduces its binding to the plasma membrane 42

3.5		Membrane surface charge influences the plasma membrane localization of MBP	46
3.6		Role of MBP-PIP2 binding for the maintenance of myelin integrity	51
3.7		A hypothetical role of MBP in regulating membrane tension	53
3.8		Characterization of oligodendrocytes during myelination in vitro	55
3.9		Polarization of oligodendrocytes	57
	3.9.1	PIP3 accumulates in Oli-neu cells at the tips of processes	57
	3.9.2	Rho inhibition correlates with PIP3 accumulation at the tips of cellular processes	59

4 Discussion 62

4.1		PIP2 dependent association of MBP to the plasma membrane	62
4.2		Possible roles for MBP-PIP2 interaction	63
	4.2.1	PIP2 as a targeting-signal to the plasma membrane	63
	4.2.2	The Role of MBP in organizing myelin lipids into microdomains	64
	4.2.3	Alteration of charges induces loss of compaction and binding of MBP	66
		4.2.3.1 Alteration of membrane charge	66
		4.2.3.2 Reduction of charges in MBP and its effect on membrane association	70
	4.2.4	Possible Involvement of MBP in process outgrowth and alteration in membrane tension	72
4.3		Conclusion	74

5 Supplemental material 75

References 79

List of Figures

1.1	Morphology of myelin and developmental stages of oligodendrocytes	3
1.2	Structure of shiverer (MBP-/-) myelin	6
1.3	PIP2 and its functions	10
3.1	Colocalization of MBP with PIP2 and PIP3 sensors	32
3.2	FRET analysis reveals close localization of MBP with PIP2	34
3.3	Mislocalization of MBP due to loss of PIP2 from the plasma membrane	36
3.4	Reduced binding of MBP due to specific reduction of PIP2 and PIP3 levels	38
3.5	PIP2 dependent plasma membrane localization of MBP	40
3.6	Localization of MBP after PIP2 depletion	43
3.7	Domains of MBP needed for plasma membrane association	44
3.8	Quantification of the membrane association of different MBP mutants	45
3.9	Decrease in surface charge influences MBP localization	47
3.10	Surface charge reduction leads to dissociation of MBP	49
3.11	Release of MBP from myelin membrane upon ionomycin treatment	50
3.12	Vesiculation of myelin due to reduction of surface charge	52
3.13	Reduced vesiculation through neomycin blockage	53
3.14	Bleb formation induced by ionomycin treatment	54
3.15	Oligodendrocyte-neuronal co-cultures stained at different developmental stages	56
3.16	PIP3 accumulates at tips of Oli-neu processes	57
3.17	Polarization signals in Oli-neu cells	58
3.18	Accumulation of polarization factors in oligodendroglial cells	60
4.1	Lateral sequestration model of PIP2 molecules	65
5.1	Biochemical quanitfication of plasma membrane localization of MBP	76
5.2	Generation of stable cell-lines expressing MBP14k-YFP and MBP21k-YFP	78

1 Introduction

During evolution, the vertebrate nervous system has developed a mechanism to insulate axonal segments. The formation of a specialized lipid enriched structure, the myelin sheath, has become the most abundant vertebrate membrane structure. Myelin is composed of multiple membrane lamellae wrapped around axons (reviewed in Trapp and Kidd; 2004). The development of such a lipid-rich structure around axons has several evolutionary advantages. Since a lipid rich membrane is non-conductive and prevents current leak, myelin exhibits an insulation of the axon. It speeds the conduction of nerve impulses by a factor of 10 compared to unmyelinated fibers. Information can thus be processes and delivered more efficiently. Additionally, the energy cost of nerve impulse propagation is reduced 100-fold. Moreover, the myelin structure and proper compaction is essential for the maintenance of axonal integrity (Fig. 1.1A, Waxman, 1997; Jessen, 2004). The loss of the myelin sheath in diseases such as Multiple Sclerosis or Leukodystrophy therefore results in a debilitating condition. Understanding the mechanisms of myelin maintenance and compaction is a prerequisite for understanding such diseases.

1.1 Composition and structure of myelin

The myelin membrane is spirally wrapped around axons and its cytoplasmic as well as exoplasmic sides are compacted through various proteins. The compacted cytosolic side of the membrane is called the major dense line (MDL) and the compacted extracellular side the intraperiod line (IPL). Non-compacted parts of myelin, which include cytosol can also be found in the myelin sheath: Compact myelin is found at the internodes, whereas non-compact myelin is found at the inner loop surrounding the axonal membrane, the outer loop, the paranodal loops, and (in the PNS) Schmidt-Lanterman incisures (Fig. 1.1A).

Although myelin can be considered as a long extension of plasma membrane, it differs from the plasma membrane in its unusual high lipid content (70% of total dry weight), its lipid composition and the accumulation of specific proteins. It is enriched in phospholipids, glycosphingolipids (in particular galactosylceramide and sulfatides), and cholesterol. The proteins found in myelin differ from other cellular compartments. Additionally, compact myelin differs from non-compact myelin in its lipid as well as its protein composition. The most abundantly expressed CNS myelin proteins are proteolipid protein (PLP) and myelin basic protein (MBP). They constitute 80% of the total myelin proteins. Apart from several glycoproteins found in myelin, such as myelin-associated protein (MAG) and myelin oligodendrocyte glycoprotein (MOG), distinct enzymes within myelin were shown to be involved in myelin turnover and maintenance (Ledeen, 1984).

Myelin has an unusual high content of glycosphingolipids, which has a tendency to partition into lipid-ordered domains. These lipid ordered domains within myelin are enriched in cholesterol and can be extracted as detergent resistant membranes (Simons et al., 2000; Lee, 2001; Baron et al., 2003; Debruin and Harauz, 2007). Most myelin lipids are transported to the myelin membrane in vesicles. The sorting of different lipids can therefore occur within the vesicular transport machinery: the endoplasmic reticulum (ER), trans-golgi network (TGN), recycling endosomes, late endosomes and the plasma membrane itself. Because of the specific lipid composition in myelin, it is likely that some of these lipids, synthesized in the ER, are preassembled early in the secretory pathway and are then transported to the newly formed myelin membrane (Maier et al., 2008). Additionally, many of the myelin proteins are also raft-associated. It is interesting to note that also within the myelin membrane, proteins can be segregated into raft associated and non-associated fractions, which are later incorporated into compact or non-compact myelin respectively. The formation of myelin can therefore be compared to the apical membrane of endothelial cells.

1.2 Oligodendrocytes

Different cell types in the peripheral (PNS) and central nervous system (CNS) form the myelin sheath; oligodendrocytes are the myelinating glia of the CNS, and Schwann cells

1 Introduction

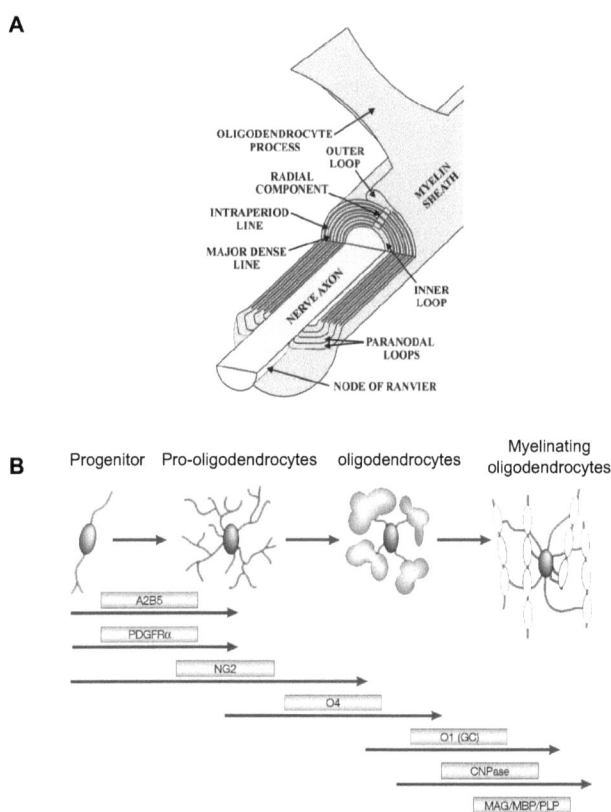

Figure 1.1: (A) Morphology of CNS myelin. Oligodendrocytes extend their processes to form the myelin sheath around axons. The myelin sheath can be segregated into compact and non-compact myelin segments (at the paranodal loops, Schmidt-Lanterman incisures, the inner− and the outer loop). Distinct proteins are involved in the formation of the major dense line and the intraperiod lines.
(B) Oligodendrocytes cultured with neurons in vitro form the myelin sheath around axons or if co-cultured without neurons extend their membranes on the coverslip. This membrane is comparable in its protein and lipid composition to myelin membrane that is formed around the axon. The different developmental stages of oligodendrocytes can be distinguished through the expression of the depicted marker antigens. Myelin proteins of the compact myelin are only expressed later in development (Images modified from Debruin and Harauz, 2007; (B) Zhang, 2001).

of the PNS. Whereas oligodendrocytes extend their processes towards different axons surrounding their cell body and myelinate up to 50 different axons, Schwann cells form a 1:1 relation with the axon and their cell soma is closely associated to the nerve fiber. Oligodendrocytes produce up to 5000 μm^2 of myelin membrane surface area per day (Pfeiffer et al., 1993). In order to study myelination, different culture systems have been developed (Lubetzki et al., 1993; Kleitman et al., 1998; Trajkovic et al., 2006; Chan et al., 2006; Taveggia et al., 2008). Oligodendrocytes can be cultured in vitro together with or in the absence of neurons (Fig. 1.1B). If oligodendrocytes are cultured in the absence of neurons, they extend their membranes on the coverslip. These membranes resemble the myelin membrane formed around axons in its lipid and protein composition and can therefore be used as a model system to study myelination. In co-culture with neurons, however, oligodendrocytes extend their processes towards axons and myelinate them. The wrapping around the axon is followed by compaction of the myelin membrane (Baumann and Pham-Dinh, 2001).

1.3 Developmental stages of oligodendrocytes

Oligodendrocyte precursor cells (OPCs) arise from the neuroepithelium of the ventricular/subvetricular zone of the developing spinal cord and brain. Oligodendrocyte progenitor cells (OPC) migrate into the developing white matter while remaining mitotic until they have reached the brain region they myelinate.

Recent in vivo time lapse imaging studies of zebra fish OPC have shown that OPCs continuously extend and retract their processes during migration into the developing white matter. During differentiation of the oligodendrocytes, these processes seem to regulate the optimal spacing between individual myelinated axonal segments (Kirby et al., 2006). Once OPCs have reached their target axon they become postmitotic and start to differentiate into myelin forming oligodendrocytes. The discrete stages of maturation of oligodendrocytes can be distinguished by the expression of different developmental markers and the apparent morphological changes (Fig. 1.1B, Jessen, 2004): oligodendrocyte type 2 astrocytes strongly express PDGF-receptor-α and sythesize gangliosides (recognized by the A2B5 antibody) (Eisenbarth et al., 1979; Hall et al., 1996). Early postmitotic oligodendrocytes can be stained for the membrane glycoprotein NG2, and are therefore often referred to as NG2 cells (Polito and Reynolds, 2005). These cells have already developed multiple processes extending towards the axons. In the adult brain NG2 cells are still present, and may serve the brain as a resource to provide new OPCs that can differentiate into oligodendrocytes. The pro-oligodendrocyte stage can be labeled with O4. The O4 antibody recognizes sulfatide (Bansal et al., 1989) and labels a cell stage relatively late in oligodendrocyte lineage. This stage is followed by the premyelinating oligodendrocyte, which is Galactocylcerebroside-positive (GalC, recognized by O1- antibody). The myelin forming oligodendrocytes generate sulfatide (recognized by O4-antibody). Amongst other marker antigens, fully differentiated oligodendrocytes express myelin proteins such as myelin basic protein (MBP) and proteolipid protein (PLP) and its splice variant DM20.

1.4 Myelin basic protein

MBP is one of the major CNS myelin proteins found in compact myelin. It is the second most abundant protein of CNS myelin after PLP. In fact, 30% of total protein and 10% of the dry weight of myelin is comprised by MBP (Boggs, 2006). Additionally, it is the only protein known so far that is absolutely necessary for myelin formation since its lack leads to a strong developmental phenotype with an almost complete loss of myelin (Roach et al., 1983; Readhead et al., 1987). MBP is a positively charged, natively unfolded protein. Natively unfolded proteins have the ability to assimilate their structure according to the environment (Uversky, 2002). If bound to ligands with opposite charged ions, natively unfolded proteins form into more structured domains. Their mean net charge is thereby reduced. Much like other natively unfolded proteins, MBP is thought to form its tertiary structure through binding with its physiological ligand, the plasma membrane (Boggs, 2006). Natively unfolded proteins have many different functions and are often involved in intracellular signaling. Various in vitro data also imply a signaling role for MBP, since it was suggested to bind to actin, microtubules, Ca^{2+}/CAM, tropomyosin, and clathrin (Grand and Perry, 1980; Modesti and Barra, 1986; Boggs and Rangaraj, 2000). Although it resembles other natively unfolded proteins such as microtubule associated protein (MAP), α-synuclein and MARCKS, unlike these proteins the positively charged amino acids of MBP are not

1 Introduction

clustered into domains. Instead, they are distributed homogenously within the amino acid sequence of MBP (Smith, 1992; Boggs, 2006). It was suggested that MBP might bind to two myelin membranes at the same time, thereby mediating compaction. Due to the number

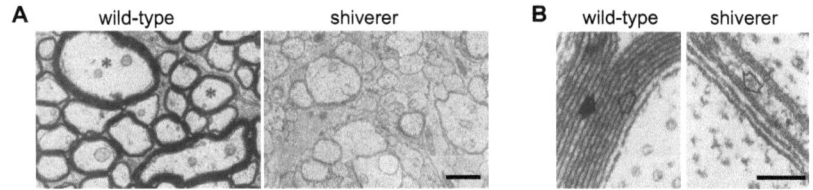

Figure 1.2: Shiverer mice lack most of the MBP gene and display a strong developmental phenotype. Due to the lack of MBP, oligdendrocytes of shiverer mice are unable to form myelin membrane around axons (A). The minimal amount of lamellae that are formed are not compacted, implicating that MBP is involved in the compaction of myelin (B). MBP is the only protein, known so far which leads to an almost complete loss of myelin. Asterisks represent the myelinated axons (Images modified from Readhead et al., 1987).

of positive amino acids, MBP has a net positive charge of 20 and at physiological pH the isoelectric point of MBP is above 10 (Rispoli et al., 2007). It is attached to the plasma membrane on the cytosolic side and it binds to acidic lipids with such strong affinity that it can only be delipidated with acid (Lowden et al., 1966; Omlin et al., 1982). It is known that it binds to the myelin membrane mainly through electrostatic interactions (Demel et al., 1973; Smith, 1977). However, hydrophobic interactions of MBP with the cell membrane due to hydrophobic or neutral amino acid stretches have also been reported (Smith, 1992; Nabet et al., 1994; Bates et al., 2003). In fact, the association of the positively charged protein MBP with negatively charged lipids, might result in membrane adhesion (Hu et al., 2004; Boggs, 2006). As a result of alternative splicing of the primary mRNA transcript, there are different splice isoforms of MBP (in mice: 21.5, 20.2, 18.5, 17.24, 17.22 and 14kDa, de Ferra et al., 1985; Campagnoni and Campagnoni, 2004). All isoforms were shown to interact with the plasma membrane. The smallest 14 kDa isoform is encoded by exons 1, 3, 5 and 7. The mRNA of MBP is targeted to the myelin membrane in granules and is translated directly at the plasma membrane. Exon 2-containing MBP isoforms (17 kDa and 21.5 kDa) were shown to accumulate in the nucleus (Pedraza et al., 1997). It is not known what role the targeting of MBP to the nucleus might play, since most of these data are based on observations in transient overexpressing cells.

1.4.1 Posttranslational modifications of MBP

The 18 kDa isoform, the most abundant in human myelin, occurs as various charge isomers. It was shown in vitro to be post-translationally modified: It is deamidated, phosphorylated, N-terminally acylated, methylated and citrullinated as well as ADP-ribosylated (reviewed in Harauz et al., 2004). These modifications lead to an alteration of the net charge of MBP and further diversification of the MBP family. The alteration of the net positive charge influences the binding affinity of MBP to acidic membranes (reviewed in Boggs, 2006). In fact, it was shown that a higher portion of citrullinated MBP is found in samples from Multiple Sclerosis patients, which underlines the role of its charge in binding to the plasma membrane (Boggs et al., 1999; Kim et al., 2003). Deiminated myelin basic protein also has a reduced ability to aggregate lipid vesicles (Harauz 2004). In vitro studies have shown that MBP is phosphorylated through protein kinase C, cAMP-dependent protein kinase (PKA), and mitogen-activated protein kinase (MAPK) family. Apart form net charge, phosphorylation might also alter the conformation of MBP. Different from methylation or citrullination, phosphorylation is reversible and might therefore also play a role in cellular signalling events. Since the posttranslational modifications mainly rely on in vitro data, their physiological relevance still remains unclear.

1.4.2 Myelin of shiverer (MBP-/-) mice

Shiverer (shiv) is an autosomal recessive mutant mouse, which shows almost complete loss of CNS myelin (Fig. 1.2). Since in shiverer mice exons 2-7 are absent, no MBP isoforms are functionally expressed (Molineaux et al., 1986). These mice therefore have provided a useful tool in studying the function of MBP. It is thought that the interaction of MBP with the cytoplasmic leaflets of the membrane bilayer causes the two opposing layers to physically associate, leading to myelin membrane compaction at the MDL (Omlin et al., 1982; Smith, 1992; Riccio et al., 2000). Corrections for the shiv myelin phenotype have been achieved by introducing the MBP gene into shiv mice. These transgenic mice produce about 25% of normal MBP protein levels and form compact myelin, indicating that one of the major functions of MBP is the compaction of myelin lamellae (Readhead et al., 1987). Since the expression of the MBP gene parallels with the process of myelination, MBP has not only been implicated in the compaction, but also the formation of myelin (Carson et al., 1983; Zeller et al., 1984). Since shiverer oligodendrocytes fail to form myelin even in adult stages, MBP has also been implicated in the generation of myelin and possibly in the targeting of myelin components. In fact, it was shown that the lack of MBP leads to a loss of lipid organization in the myelin membrane (Fitzner et al., 2006; Hu and Israelachvili, 2008). MBP accumulates in detergent-resistant membranes. Proteins incorporated into myelin such as MBP can therefore be isolated from the brain through detergents such as TritonX-100 or CHAPS as detergent resistant membranes (Debruin and Harauz, 2007). Additionally, experimental data suggest that neuronal signals induce this specific targeting of MBP into detergent resistant membranes (Fitzner et al., 2006). Still unsolved is how MBP leads to the compaction of myelin, how MBP interacts with the myelin membrane and how it can lead to the clustering of lipids in cells. Although shiverer mice have been extensively studied, it still remains elusive, why shiverer oligodendrocytes fail to function.

1.5 The phosphoinositide PIP2

Compared to other phospholipids, PIP2 is a highly negatively charged lipid. At physiological pH, it has a valence of -4. Apart from PIP3, other phospholipids only have a valence of -1. Additionally, different from other lipids, the valence of PIP2 depends on many factors, such as local pH, proteins binding to it and local ion concentration (McLaughlin et al., 2002). In Fig. 1.3A the structure of PIP2 is illustrated. It has been suggested that due to its charge and structure, PIP2 penetrates further into the aqueous phase than other phospholipids (McLaughlin et al., 2002). The concentration of PIP2 within a cell was calculated as $10\,\mu$M, which is equivalent to 1% of phospholipids (Gamper and Shapiro, 2007b).

1.5.1 Functions of PIP2

PIP2 regulates many different cellular processes: exo- and endocytosis, membrane trafficking, protein trafficking, phagocytosis, activation of enzymes, receptors and channels (Fig. 1.3, McLaughlin et al., 2002; McLaughlin and Murray, 2005). It serves as a second messenger precursor, giving rise to IP3, DAG and also PIP3. These molecules are present in a low concentration in quiescent cells and can increase in concentration upon receptor activation. They are therefore ideal as second messengers. For example, the binding of phospholipase Cδ1 (PLCδ1) to PIP2 localizes it to the plasma membrane. Upon receptor activation, PLC is activated and hydrolyses PIP2 to IP3 and DAG, which leads to a rise in intracellular Ca^{2+}. Increased intracellular Ca^{2+} concentration results in activation of various enzymes. Thereby, PLCδ1 is not activated through its binding to PIP2 but through binding to Ca^{2+}. The membrane anchorage through the binding to PIP2 simply facilitates hydrolysis. Furthermore, a direct link between PIP2 and proteins that bind to cytoskeleton has been shown to influence the membrane tension and therefore the shape of cells (Raucher et al., 2000). Additionally, PIP2 was shown to be involved in exocytosis and clathrin-mediated endocytosis. Various studies have indicated that PIP2 is involved in the insertion and uptake of membrane and is therefore important in regulating plasma membrane morphology (Mellman, 2000; Golub and Caroni, 2005). Membrane trafficking can therefore influence the cell shape and the composition of the plasma membrane. Taken together, PIP2 is thought to organize membrane extension and overall cell shape. PIP2 was also shown to bind scaffolding proteins, and is involved in the regulation of ion channels (Hilgemann et al., 2001). Additionally, during phagocytosis, PIP2 is concentrated in nascent phagosome and membrane ruffles and was shown to play a role in initial cup formation (Yeung et al., 2006a). Recent findings have shown that PIP2 and PIP3 play a crucial role in apical membrane formation during epithelial cyst formation and axon specification (Martin-Belmonte et al., 2007). PIP3 and its precursor PIP2 might also play a major role in the polarization of oligodendrocytes. During development, oligodendrocytes form processes that are later retracted. Oligodendrocyte polarization is therefore comparable to neuronal polarization (Simons and Trotter, 2007). In neurons several initial processes are formed, before one of them receives a positive signal to extend, thereby sending a retractive signal to other processes. During the formation of the initial axon, the polarization-inducing complex, composed of mPar3/mPar6/PKC/APC,

1 Introduction

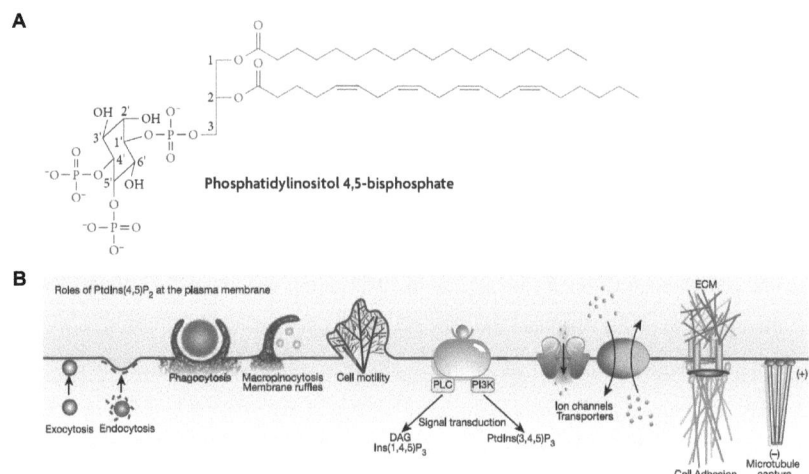

Figure 1.3: (A) Molecular structure of PIP2. PIP2 is formed through phosphorylation of 4' and 5' OH groups within the inositol ring mainly at the plasma membrane through PIPK type I activity. Thereby the overall charge of this phospholipid is reduced to -4 at a physiological pH. PIP2 is anchored to the membrane through two poly-unsaturated hydrocarbon chains.
(B) PIP2 is involved in almost all cellular processes. It regulates membrane extension and cell shape (through regulation of exo- and endocytosis, phagocytosis, membrane ruffles, cell motility, cell adhesion and it is involved in the capture of microtubules). It also plays a role in signal transduction pathways and serves as a second messenger precursor (IP3 and PIP3) as well as an activator of ion channels or receptors (Images modified from Di Paolo and De Camilli, 2006).

drives specification of the axon (Shi et al., 2003; Banker, 2003; Martin-Belmonte et al., 2007; Goldstein and Macara, 2007). It was shown that PIP3 plays a crucial role in myelin formation in the central nervous system (Flores et al., 2008). Loss of the PIP3-phosphatase PTEN in oligodendrocytes of mutant mice leads to hypermyelination, indicating that an increase of PIP3 drives myelin formation. An increase of PIP3 in oligodendrocytes results in increased myelination, even in adult mice (Goebbels et al., unpublished observation).

1.5.2 Enzymes generating PIP2

Phosphorylation of different OH groups within the inositol ring through different PIPkinases leads to the generation of distinct phosphoinositols. The enzymes are located at distinct subcellular localizations (Anderson et al., 1999). Phospholipids are therefore inhomogenously distributed within the membrane system of the cells (Krauss and Haucke, 2007). PIP2 is mainly generated upon phosphorylation of PI4P at the 5' OH-group through the PIPK typeI and can be found at the plasma membrane. The PIP2 kinases are activated upon receptor stimulation and can lead to localized production of PIP2 (Doughman et al., 2003). The small GTPase Arf6 is also involved in PIP2 production, by activating PI4P5K, and induces the accumulation of PIP2 at the plasma membrane (Donaldson, 2003). The main phosphatases of PIP2 are Synaptojanin 1 and SHIP1. Synaptojanin 1 is involved in the regulation of PIP2 levels and is indispensable for PIP2 recycling during vesicle trafficking (Milosevic et al., 2005). Synaptojanin 1 knockout mice therefore display a developmental phenotype and fail to develop, mainly due to failure in endocytosis (Cremona et al., 1999). PIP2 is thought to be more abundant in myelin compared to other membranes of other cell types. Studies of ^{32}P incorporation into myelin have shown, that within 60 min of incubation about 15% of ^{32}P-labeled PIP2 is incorporated into myelin (Deshmukh et al., 1981; Kahn and Morell, 1988). Additionally it is believed that PO^{3-} groups are provided by the axon. It is therefore not surprising that all PIP2 generating enzymes can be found in the myelin sheath (Chakraborty et al., 1999).

1 Introduction

1.5.3 Protein domains binding to PIP2

There are several known protein domains that can bind PIP2: unstructured-domains, tubby, pleckstrin homology (PH); phox homology (PX); epsin N-terminal homology (ENTH) domains, four-point-one, ezrin, radixin, moesin (FERM) and many other. PH-domains bind with high affinity to PIP2, the PH-domain of PLCδ1 for example binds with $K_d = 2\mu M$. It comprises 120 amino acids and binds to PIP2 in a 1:1 ratio (McLaughlin et al., 2002). The binding of the unstructured domain (also called natively unfolded proteins) of 'myristoylated alanine-rich C kinase substrate' (MARCKS) to PIP2 has been under intensive investigation (reviewed in Sheetz et al., 2006). MARCKS binds several PIP2 molecules at the same time mainly through electrostatic interaction of distinct Lys residues. MARCKS was also shown to bind to Ca^{2+}/CAM, PKC and actin. Phosphorylation through PKC and binding of Ca^{2+}/CAM was shown to influence the binding of MARCKS to the plasma membrane. Several other unstructured domains were reported to bind to PIP2. Among them are actin-binding proteins, such as GMC (GAP43, MARCKS, CAP23), also referred to as PIPmodulins due to their ability to cluster PIP2 at the cell membrane (Laux et al., 2000).

1.5.4 Molecular tools to monitor phospholipids

Since recent years it has become apparent that phosphoinositides take part in the control of almost every aspect of a cell's life. Recent findings have led to the demand of new technologies to visualize PIP2, in order to study the spatio-temporal aspects of inositide signaling. Since PH domains of various proteins bind with different affinity to distinct phosphoinositides, these lipids can be visualized by adding a fluorescent tag to these PH domains. While the PH domain of AKT has widely been used to visualize PIP3, the PH domain of PLCδ1 has been used to visualize PIP2. Additionally, FRET pairs of different PH domains have been used to study the activity of PLC (van der Wal et al., 2001). Upon PLC activation, PIP2 is hydrolyzed to IP3 and DAG, which results in the dissociation of GFP-PH-PLCδ1. Cotransfection of the FRET pair composed of PH-PLCδ1 with a CFP and YFP tag respectively therefore showed a decreased FRET efficiency upon PLC activation. These fluorescently tagged domains are therefore widely used to study dynamics and interactions of phosphoinositides (van der Wal et al., 2001; Várnai and Balla, 2007). In this study I used different sensors to monitor the distribution and dynamics of phosphoinositides.

2 Materials and Methods

2.1 Materials

Chemical reagent were purchased from SIGMA, unless noted otherwise.

2.1.1 Cell Culture

2.1.1.1 Mammalian cell lines

COS-1	Green monkey kidney, fibroblast cells
Oli-neu	Rat, O2A cells
OLN-93	Rat, O2A cells
BHK	Baby hamster kidney cells

2.1.1.2 Mammalian cell culture media

DMEM for mammalian cell culture was purchased from GIBCO or BioWhittaker, BME and OptiMEM from GIBCO.

BHK cell medium
2.5% Horse serum
OptiMEM
10% Tryptose phosphate broth
Hepes
Penicillin/Streptomycin (BioWhittaker)

OLN-93 cell medium
DMEM
10% FCS
Penicillin/streptomycin

Oli-neu cell medium (SATO)

Concentration	Component
10 μg/ml	Insulin
1 μg/ml	Transferrin
25 μg/ml	Gentamycin
220 nM	Sodium-Selenite
520 nM	L-Thyroxine
500 pM	Tri-iodo-threonine
100 μM	Putrescine
200 nM	Progesterone

DMEM 4,5 g/l glucose
Sterile filter and add between 1 to 5 % Horse Serum

Freezing Medium for all cell lines
70% DMEM
20% FCS (PAN Biotech)
10% DMSO

Primary oligodendrocyte medium
A. before shake
BME
10% HS
Penicillin/Streptomycin
B. after shake
Sato medium 1% HS

2.1.2 Strains and cells

2.1.2.1 Bacterial strains

Escheria Coli
  DH 5α
  XL1-Blue

2.1.2.2 Bacterial culture media

Prior to the use, bacterial media were autoclaved and supplemented with antibiotics.

LB-Medium
  1 %   Bacto Tryptone
  0.5 %   Bacto Yeast extract
  1 %   NaCl

Make 1000 ml with H_2O, set pH 7.5 with 10 N NaOH and autoclave.

Antibiotics were used at the following concentrations:
  150 mg/l   Ampicillin
  25 mg/l   Kanamycin
  50 mg/l   Chloramphenicol

2.1.3 Molecular cloning reagents

2.1.3.1 Plasmids

pEYFP-N1	BD-Biosciences Clontech
pECFP-N1	BD-Biosciences Clontech
pGEMT	Promega
pEYFP-Mem	Clontech
pMSCVhyg	BD-Biosciences Clontech
pET22b(+)	Novagen
pSFVgen	provided by M. Simons

2.1.3.2 Enzymes

Pfu	Stratagene
Dpn1	NewEngland Biolabs
Taq	SIGMA
Easy-A DNA	Stratagene
T4 ligase	Promega

2.1.3.3 Buffers

DNA-sample buffer (6x)
20%(w/v)   Glycerol in TAE buffer
0.025% (w/v)   Orange G bromphenol blue
dNTP   stock solution (100 nM)
25 mM   each dATP, dCTP, dGTP, dTTP(Boehringer, Mannheim)
1μg/ml Ethidiumbromide for agarose gels in TAE

TAE (50x, 1000ml)
2 M   Tris-Acetate, pH 8.0
50 mM   EDTA
57.1 ml   Glacial acetic acid
make 1000 ml with ddH$_2$O

2.1.3.4 Primer sequences and PCR protocol

Primer	Sequence
K5A sense	CACCATGGCATCACAGGCGAGACCCTCACAGCGA
Cluster 1 K5A antisense	TCGCTGTGAGGGTCTCGCCTGTGATGCCATGGTG
R6A sense	GCCACCATGGCATCACAGGCGGCACCCTCACAGCGATCCAA
R6A antisense	TTGGATCGCTGTGAGGGTGCCGCCTGTGATGCCATGGTGGC
Cluster 2 R10A/K12A sense	CAGAAGAGACCCTCACAGGCATCCGCGTACCTGGCCACAGCAAG
antisense	CTTGCTGTGGCCAGGTACGCGGATGCCTGTGAGGGTCTCTTCTG
Cluster3 R24A sense	CCATGGACCATGCCGCGCATGGCTTCCTCC
R24A antisense	GGAGGAAGCCATGCGCGGCATGGTCCATGG
H22A sense	CACAGCAAGTACCATGGACGCTGCCGCGGCTGGCTTCCTCCCAAGGCAC
H22A antisense	GTGCCTTGGGAGGAAGCCAGCCGCGGCAGCGTCCATGGTACTTGCTGTG
Cluster 4 R30A/R32A sense	GCATGGCTTCCTCCCAGCGCACGCAGACACGGGCATCC
R30A/R32A antisense	GGATGCCCGTGTCTGCGTGCGCTGGGAGGAAGCCATGC
Cluster 5 sense	ACAGGGTGCGCCCGCGGCGGGCTCTGGCAAGG
antisense	CCTTGCCAGAGCCCGCCGCGGGCGCACCCCTGT
Cluster 6+7 sense	GCATCCTTGACTCCATCGGGGCCTTCTTTAGCGGTGACA
antisense	TGTCACCGCTAAAGAAGGCCCCGATGGAGTCAAGGATGC
exon 1 sense	CCG GAA TTC GCC ACC ATG GCA TCA CAG AAG AGA
exon 1 antisense	CGC GGA TCC TTG CCA GAG CCC CGC TT
Exon 1 MBP14k-YFPS54A sense	CCC AAG CGG GGC GCT GGC AAG GAT C
Exon 1 MBP14k-YFPS54A antisense	GGG TTC GCC CCG CGA CCG TTC CTA G
MBP14k-YFP full length S54A	CCC AAG CGG GGC GCT GGC AAG GAC TCA CAC ACG AG
MBP14k-YFP full length S54A	CTC GTG TTG TGA GTC CTT GCC AGC GCC CCG CTT GGG

PCR

2.5 μl	sense primer
2.5 μl	antisense primer
50-200ng	cDNA template
0.4 μl	PfuI
2 μl	10x Pfu buffer
2 μl	dNTP mix
10.3 μl	ddH$_2$O

1. 95 °C   3 min (denaturation)
2. 95 °C   30 s (denaturation)
3. 55 – 60 °C   60 s (annealing, dependent on T_m)
4. 68 °C   6 min (amplification, 30 cycles to step 2)
5. 68 °C   10 min
6. 4 °C   pause

DpnI digest

20 μl	PCR mix
1 μl	DpnI
5 μl	Buffer 4 (NEB)
24 μl	ddH$_2$O

2.1.4 Biochemical reagents

2.1.4.1 Western Blotting reagents

Blocking Buffer
5 % non fat dry milk powder in TBS

Blotting buffer
1x (pH unadjusted, Western Blotting)
39 mM   Tris-HCl
48 mM   Glycine
10 % (w/v)   Methanol

SDS sample buffer (5x)
10% (w/v)   SDS
10 mM   Dithiothreitol
20% (v/v)   Glycerol
0.2 M   Tris-HCl, pH 6.8
0.005% (w/v)   Bromphenolblue

SDS running buffer (1x)
25 mM   TrisHCl
192 mM   Glycin
1% (w/v) SDS

SDS separating gel
12.0 % (1 gel of 1.5mm thickness)
4 ml   30% polyacrylamid (BioRad)
10 ml   separation gel buffer
         (1.5 M Tris-HCl; 0.4% (w/v) SDS), pH 8.8
3.5 m l ddH$_2$O

SDS stacking gel
12% (1 gel of 1.5mm thickness)
0.8 ml   30% poly-acrylamid (BioRAD)
1.5 ml   stacking gel buffer (0.5M Tris-HCl
0.4%(w/v)   SDS, pH 6.8
3.7 ml   ddH$_2$O
20 µl   Ammonium persulfate (10% w/v)
20 µl   TEMED

2.1.4.2 Membrane isolation buffers

TE (1x)
10 mM   Tris-HCl, pH 8.0
1 mM   EDTA

Sucrose gradient buffer
250 mM   saccarose
3 mM   imidazol
(add protease inhibitor tablets fresh)

TNE
10 mM   Tris-HCl, pH 7.4
0.2 M   NaCl
1 mM   EDTA

2.1.5 Immunofluorescence labeling reagents

Immunocytochemistry buffers

PBS 1x, cell culture
136 mM   NaCl
2.6 mM   KCl
10 mM $Na_2HPO_4 \times 2H_2O$
1.4 mM KH_2PO_4
make 1000 ml with ddH_2O,
set pH to 7.2 with 10 N NaOH

TBS
25 mM   Tris-HCL, pH 7.5
136 mM   NaCl
2.6 mM   KCl

Fixative
4% Paraformaldehyde in PBS/TBS

Mounting Agent
Aqua poly-mount (Polysciences)

Blocking solution
BME
10%(w/v) Horse serum

2.1.6 cDNA Clones

mRFP-LactC2	provided by S. Grinstein, Toronto, Canada
GFP-PH-PLCδ1	
GFP-PH-PLCδ1-3xmut	obtained from I. Milosevic and J. Soerensen (MPI
IPPCAAX-GFP-SFV(mRFP-Synj1)	for Biophysical Chemistry, Göttingen, Germany)
PIP4P5K-GFP-SFV	
ARF6/Q67L-HA	provided by J.G. Donaldson, NIH, Bethesda, Maryland
PH-AKT-YFP	provided by T. Meyer (Stanford, CA)
MBP cDNA	provided by T. Campagnoni (UCLA, CA)
ΔExon1(MBP)-YFP	
MBP14k-YFP	provided by Angelika Kippert (University of
MBP21k-YFP	Göttingen; department of Biochemistry II)
ΔExon(1,3,5)-YFP	
Exon7-YFP	

2.1.7 Antibodies

NG2	mouse	provided by J Trotter	1:50
MBP	rabbit	DAKO	1:200
MBP	mouse	DAKO	1:200
Tuj1	mouse	Covance	1:500
Lamp1	mouse	PharMingen	1:100
KDEL (BIP)	mouse	Stressgen Biotechnologies	1:100
GM130	mouse	BD Transduction Laboratories	1:100
HA	rat	production within Nave lab	1:50
Cy5	mouse	Chemicon	1:250
cy3	rabbit	Dianova	1:4000
cy3	mouse	Dianova	1:2000
cy2	mouse	Dianova	1:2000
O1(anti-GalC)	mouse	provided by J Trotter	1:50
O4(anti-sulfatide)	mouse	provided by J Trotter	1:50
Par3	rabbit	provided by T Pawson	1:100
phospho-AKT	rabbit	Cell Signalling	1:200

2.1.8 Chemical compounds

Ionomycin	- dissolved in DMSO, Calbiochem
Wortmannin	- dissolved in DMSO, Sigma
DMSO	- Sigma
Antimycin	- Sigma
2-deoxy-D glucose	- Sigma
Neomycin (G418)	- Invitrogen

Chemical reagents for electron microscopy

Embedding medium
compounds all from Serva
- 21.4 g   Glyconerin ether, 1,2,3- propanetriol glycindyl ether
- 14.4 g   2-dodecenylsuccinic acid anhydride
- 11.3 g   methylnadic anhydride
- 0.84 g   2,4,6-tri(dimethylaminoethyl)phenol

— 4% uranyl acetat
— luxol fast blue

2.2 Methods

2.2.1 Manipulation of DNA

For restriction digestion with type II endonucleases 1 µg DNA was incubated with 5 to 10 units of Enzyme for required times and at required temperatures. Restriction was terminated by heat inactivation. Ligation of DNA fragments was performed by mixing 1-3 ng of DNA with three-fold molar mass of insert with T4 Ligase and 2 µl of 10x ligation buffer and added with double distilled water to final volume of 20 µl. Ligation was incubated for 2 h at room temperature. Ligation mix was used directly for transformation.

2.2.2 Transformation

An aliquot of competent cells was thawed on ice. 1.7 µl of β-mercaptoethanol (of 1:10 dilution) was added to the cells and incubated for 10 min on ice. Plasmid was added (5 µl of ligation mix and 25 µl of PCR product) and incubated for additional 30 min on ice. After cells were kept on ice, a short (40 s) heat pulse at 42 °C was given. Then cells were kept for another 2 min on ice before suspending them in prewamed LB-medium (without antibiotics). Cells were then incubated at 37 °C for 45 min before plating them onto LB-plates containing the appropriate resistance. Plates were incubated at 37 °C overnight single colonies were selected with sterile toothpick and incubated for 16 h in 3-4 ml LB medium.

2.2.3 Mini/Midi plasmid purification

After single colonies were incubated at 37 °C for 6-12 h in 3-4 ml LB medium containing the appropriate antibiotics, cells were centrifuged for 1 min at 900rpm with table-top centrifuge. Plasmids were isolation from bacteria according to the manufacturers protocol (Mascherey-Nagel). DNA was eluted from columns with TE (pH 8.0).

2.2.4 Generation of stable cell lines

For the generation of stable cell lines, an early passage of Oli-neu cells were plated onto poly-L lysine coated 6 well plate. Cells were co-transfected with low amounts of linearized DNA (1 μg plasmid DNA of interest and 100 ng hygromycin resistance). After 16-24 h expression was monitored and 400 μg/ml Hygromycin B was added to the culture medium. Within 10-12 days in culture, all cells without hygromycin resistance died and single colonies formed, expressing MBP14k-YFP and MBP21k-YFP. Within 20-25 days in culture signle colonies were picked with sterile pipette and plated into 96 well plate. Single clones were allowed to grow until confluent and plated into 6 well plate after. Transfection was monitored by fluorescence microscopy and western blot analyis.

2.2.5 Biochemical techniques

2.2.5.1 Sucrose gradient centrifugation

Transfected Oli-neu cells were scraped off the culture plate, which was kept on ice. Cells were homogenized with 23G needle in 500 μl sucrose buffer. Cell-homogenate was centrifuged with table-top centrifuge at 3000 rpm for 10 min at 4 °C to remove post-nuclear membrane fractions. Supernatant was centrifuged at 100000g for 1 h and the pellet was resuspended in 500 μl.

2.2.5.2 Detergent resistant membrane isolation from Oli-ceu cells

Cells were scraped off in TNE and homogenized using 27G needle. Post-nuclear membranes were removed by centrifuging samples at 4 °C for 10 min at 3000 rpm. Supernatant was collected and centrifuged with TLA120.1 rotor (Beckmann) at 100000 g for 1 h or at 13000 rpm with table-top centrifuge for 30 min. The pellet was resuspended in the same volume as supernatant. For Optiprep (Sigma) gradient centrifugation, cell homogenate was pre-incubated with 20 mM CHAPS (Biomol) before centrifugation. Samples were then loaded onto Optiprep gradient (25 μl lysate + 250 μl Optiprep, 1.2 ml Optiprep/TNE (1:1 ratio), 200 μl TNE). Optiprep gradient was centrifuged at 5000 rpm in TLS 55.2 for 2 h at 44 °C.

six fractions were taken and loaded on SDS gel for western blotting.

2.2.6 Cell culture and transfections

2.2.6.1 Primary cell culture

Primary oligodendrocytes were prepared as described (Trajkovic et al., 2006). E15 or P0 mice were sacrificed and total mouse brain was removed. The meninges were removed and incubated in 0.5% Trypsin-EDTA (Gibco) for 10 min at 37 °C in falcon tubes. Brains were washed briefly with HBSS and 10 ml of medium (containing BME, 10% (w/v) horse serum 1% (w/v) Glutamate, 1% (w/v) Penicillin/Streptomycin) was added to stop the reaction. Brains were then homogenized with 10 ml pipette and centrifuged at 800 rpm for 10 min. For oligodendrocyte-neuronal co-cultures, 500000 cells were plated onto 15 cm poly-L lysine coated coverslips. For oligodendocytic cultures, cell suspension was plated into 10 ml cell-culture flasks (Nunc) each containing cell suspension derived from 3-4 brains. Mixed cultures were then incubated at 37 °C, 5% CO_2 for 14 days before shaking off oligodendrocytes from astrocytic layer. After shaking oligodendrocytes from a this monolayer, cells were plated onto poly-L lysine-coated coverslips and cultured in DMEM with B27 supplement and 1% horse serum, L-thyroxine, tri-iodo-thyronine, glucose, glutamine, gentamycine, pyruvate, and bicarbonate (for concentrations see Materials, SATO buffer).

2.2.6.2 Oligodendroglial cell lines

The oligodendroglial precursor cell line, Oli-neu (provided by J. Trotter, University of Mainz, Germany), and OLN-93 cells (provided by C. Richter-Landsberg, University of Oldenburg, Germany) were cultured as described (Richter-Landsberg and Heinrich, 1996; Jung et al., 1995). Oli-neu cells were cultured in SATO medium containing 1-5% (w/v) horse serum. Oli-neu cells were passaged by washing off cell layer from culture dish and replating them in fresh medium onto poly-L lysine coated culture dishes. OLN-93 cells were passaged by incubating them for 2 min with Trypsin/PBS. Culture medium was then added to stop trypsin reaction and cells were centrifuged for 5 min at 800 rpm. Transient transfections were performed using FuGENE transfection reagent (Roche) or Lipofectamine transfection reagent (Invitrogen) according to the manufacturers protocol. Transient transfection of COS1 cells was performed using Lipofectamine reagent according to the manufacturers protocol. BHK cells were passaged like OLN-93 cells. To wash cells, they were centrifuged at 1200 rpm for 2 min and replated into 10 ml culture flasks (Nunc).

2.2.7 Expression constructs and virus generation

An Exon1-MBP-EYFP fusion protein was generated by cloning the PCR product of MBP exon 1 into pEYFP-N1(BD Biosciences Clontech) using EcoRI-BamHI sites. The GFP-PH-PLCδ1 fusion construct was subcloned into pECFP-N1 (BD Biosciences) using Age I-Not I sites. Doubly-palmitoylated YFPmem vector was purchased from Clontech Laboratories. All site-directed mutants were generated by circular amplification with Pfu Turbo DNA polymerase (Stratagene, La Jolla, CA) followed by digestion of methylated and hemimethylated DNA with DpnI (New England Biolabs, Ipswich, MA). All constructs were verified by DNA sequencing. Recombinant virus was generated as described (Fitzner et al., 2006). Viral RNA was generated by in vitro transcription of linearized vector plasmid and pSFV-Helper1 plasmid (linearized with SpeI). SFV-Helper RNA and the respective RNA constructs were electroporated into Baby-Hamster Kidney (BHK21) packaging cells, that were cultured at 37 °C and 5% CO_2. About 24 h after electroporation supernatant containing virus particles was collected. For transduction, primary oligodendrocytes were incubated with supernatant containing viral particles (for 1 h), before adding back the culture medium. The infection was allowed to continue for additional 7 h.

2.2.8 Immunofluorescence staining procedure

Immunofluorescence was performed as described before (Trajkovic et al., 2006). Briefly, cells were fixed with 4% PFA. For labeling of intracellular proteins, cells were permeabilized with 0.005 - 0.01% Triton X-100 (Serva) in PBS, followed by incubation with blocking solution, containing BME supplemented with 10% HS (for 30 min at room temperature). Cells were incubated with primary antibodies diluted in blocking solution for 1-4 h at room temperature, washed with PBS, and incubated with the respective secondary antibody for additional 45 min.

2.2.9 Life cell imaging and image analysis

PLC activation experiments were performed as described (Várnai and Balla, 1998). Cells were washed twice with modified Krebs-Ringer solution (containing: 120 mM NaCl, 4.7 mM KCl, 1.2 mM $CaCl_2$, 0.7 mM $MgSO_4$, 10 mM Glucose, 10 mM Na-Hepes, pH 7.4) before imaging. Coverslips were then placed into a chamber that was mounted on a heat stage and kept at 33 °C (temp control 37-2 digital, Zeiss, Germany) during image acquisition. Cells were imaged in modified Krebs-Ringer solution and fluorescent images were acquired under oil with an inverse microscope (Leica Axiovert 200M). Images were obtained using AxioVision Software at multiple x-y position with a high-resolution digital camera with a progressive scan interline CCD chip camera (ORCA ER, C4742-80-12AG; Hamamatsu, Japan). Images were acquired every 10 s using the appropriate filters (excitation filter for YFP and GFP: BP 450-490 and emission filter BP 515-565; Carl Zeiss, Germany).

For calcium entry, 10 μM ionomycin (Calbiochem) was added to the imaging solution, by removing 0.5 ml of 2 ml medium and adding back 0.5 ml medium containing reagents. For PIP2 blockage, cells were incubated for 10 min at 37 °C with 10 mM neomycin (G418, Invitrogen). After cells were treated for 2 min with ionomycin, 5 mM EGTA was added for 30 min. In all conditions, cells were fixed and mounted in aqua polymount (Polysciences) and fluorescent images were analyzed using a modified ImageJ macro. Line scans were taken directly at the cell membrane (n > 10; pixel size 100x100 nm). Statistical analysis was performed using GraphPrism software. For quantification, line scans were taken from confocal images (n > 30, average pixel size 40x40 nm; Fig. 3.8, 3.9 and 3.10).

2.2.10 Generation of membrane sheets

Sheets were generated as described (Milosevic et al., 2005). Briefly, OLN-93 cells were plated onto 25 mm poly-L lysine coated coverslips and kept at 37 °C, 5% CO_2 for 8 h, before transfection of plasmid DNA using Lipofectamine reagent. Membrane sheets were generated from OLN-93 cells 12 h after transfection. For wortmannin treatment, cells were treated with 30 nM wortmannin (Sigma) for 4 h before generation of sheets. Coverslips were placed into ice cold sonication buffer, containing 120 mM K-Glu, 20 mM K-Acetate, 20 mM Hepes, 10 mM EGTA in a total volume of 300 ml. Coverslips containing cells were placed 3 cm above the sonication tip and one single sonication pulse was applied (Sonifier 450, power setting at 2.5, duty cycle 300 ms; Branson Ultrasonics, Danbury, CT).

Sheets were fixed for 1 h in 4 % PFA and washed three times with PBS before imaging. Coverslips containing sheets were then placed into the microscope chamber. To identify intact sheets, the phospholipid bilayer was visualized by adding 1-(4-trimethyl-amoniumphenyl)-6-phenyl-1,3,5-hexatriene (TMA-DPH, Molecular probes, Eugene, OR) to the imaging solution. For imaging we used an Axiovert 100 TV fluorescence microscope (Zeiss, Oberkochen, Germany) equipped with a 100x, 1.4 numerical aperture plan achromate objective using appropriate fluorescence filter sets (excitation filter G 365, BS 395 and emission filter LP 420 were used for TMA-DPH dye; excitation filter BP 480/40, BS 505 and emission filter BP 527/30 were used for YFP). The focal position of the objective was controlled using a low-voltage piezo translater driver and a linear variable transformer displacement controller (Physik Instrumente, Waldbronn, Germany). Images were acquired using a back-illuminated CCD camera (512x512-chip with 24x24 μm pixel size with a magnifying lens; 2.5x Optovar), to avoid spatial undersampling by the larger pixels. The focal plane was adjusted by using small fluorescent beads as a reference (0.2 μm Tetraspek-beads, Molecular Probes), applied to the imaging solution. Digital images were obtained and analyzed using MetaMorph software (Universal Imaging, West Chester, PA).

For quantification, a randomly selected region of interest (ROI) was defined on the sheet and the fluorescence intensity in that ROI was compared with background on the coverslip. For each condition, over 100 sheets were measured that were taken from at least three independent experiments. Statistical significance was determined using non-parametric students t-test in GraphPrism.

2.2.11 Ionomycin treatment of primary oligodendrocytes

Primary oligodendrocytes were cultured in vitro for five days until myelin membrane sheet was developed. For ionomycin treatment cells were washed and processed in Ca^{2+}-free

Krebs-Ringer solution (see section 'life cell imaging'). Cells were first washed twice in buffer containing Ca^{2+} for ionomycin treated cells and then incubated for 2 min with 10 μM ionomycin or DMSO for control cells. All cells were then washed three times with Ca^{2+}-free buffer before treating them with 0.005% saponin for additional 1 min. Cells were then washed again with Ca^{2+}-free buffer and fixed with 4% PFA for 8 min. Cells were then immunolabeled against GalC (O1) and MBP.

2.2.12 FRET measurement

Oli-neu cells were transiently transfected with CFP-PH-PLCδ1 and MBP14k-YFP plasmids (mixed in a 1:1 ratio) using FuGENE transfection reagent. Cells were fixed with 4% PFA 12 h after transfection, and mounted on glass microscope slides in aqua polymount (Polysciences). Fluorescence images were acquired with a Leica DMRXA microscope (Leica, Germany).

FRET was detected by an increase in donor fluorescence after photobleaching of the acceptor using a confocal Leica microscope (TCS SP2 equipped with AOBS) as described previously (Fitzner et al., 2006). Acceptor photobleaching was performed using Leica microsystems software. YFP was excited at 514 nm and CFP at 458 nm HeNe laser line. Image analysis of FRET data was performed using custom-written MATLAB routines. Fluorescence emission was collected in spectral windows (collected at 525-610 nm for YFP and collected at 468-495 nm for CFP). YFP was bleached in a ROI minimal of its initial fluorescence intensity, representing background signal. All settings were kept constant for all images acquired. FRET was calculated on a pixel to pixel basis.

2.2.13 Quantification of protein localization at the plasma membrane

Oli-neu cells were transiently transfected with plasmids MBP14k-YFP or GFP-PH-PLCδ1 using FuGENE transfection reagent and incubated at 37 °C, 5 % CO_2 for 12 h. For PI3K inhibition, cells were treated for 4 h with 30 nM wortmannin. Subsequently, cells were fixed with 4 % PFA and stained against plasma membrane localized NG2 and secondary cy5 antibody (1:250, Chemicon). Fluorescent images were acquired using a Zeiss (Jena, Germany)

2 Materials and Methods

LSM 510 confocal microscope with a 63x oil plan-apochromat objective (numerical aperture NA 1.4).

NG2 staining was used as mask to calculate the intensity at plasma membrane (PM) compared to the cytosol (cyt). For this, an optimal threshold for the mask was calculated and the images of the mask were then converted to black and white. PM localization was determined by multiplying the mask with the YFP channel (corresponding to MBP14k-YFP or YFPmem signal). The PM association constant was then calculated according to the following formula (Heo et al., 2006):

$$c^{PM} = \frac{I^{before}(PM)}{I^{before}(cyt)} \frac{I^{after}(cyt)}{I^{after}(PM)}$$

with I^{before} and I^{after} as the fluorescence intensities before and after PIP2 and PIP3 reduction. For each condition, 70 cells were analyzed using custom-written MATLAB routines. Statistical significance was determined using non-parametric students t-test.

2.2.14 Acute slices of corpus callosum

30 days old mice were sacrificed and the frontal lobes were isolated. Coronal slices were cut using a Leica VT1200S Microtome (Leica, Germany) at 300 μm thickness in ice cold cutting solution containing: 130 mM NaCl, 3.5 mM KCl, 10 mM $MgSO_4$, 0.5 mM $CaCl_2$, 1.25 mM NaH_2PO_4, 24 mM $NaHCO_3$, 10 mM glucose and was maintained at pH 7.4 in 5% CO_2 atmosphere. Before treatment slices were equilibrated in ACSF containing: 130 mM NaCl, 3.5 mM KCl, 1.5 mM $MgSO_4$, 2 mM $CaCl_2$, 1.25 mM NaH_2PO_4, 24 mM $NaHCO_3$, 10 mM glucose for 2 h at RT. ACSF was continuously bubbled with carbogen (95% O_2 and 5% CO_2) gas. Slices were then incubated at 35 °C, with ionomycin (10 μM). Control slices were incubated in 1% DMSO. For PIP2 blockage through neomycin, acute slices were incubated for 15 min prior to ionomycin treatment (for additional 30 min). During this time neomcin was kept in the bathing solution. For ATP depletion, acute slices were incubated in Ca^{2+} and glucose free ACSF, containing antimycin (200 nM) and 2-deoxy-D-glucose (10 mM) for 1 h. Acute slices were then fixed for electron microscopy in Karlsson-Schultz fixative for 8 h, containing: 2.5% glutaraldehyde, 2% formaldehyde, 0.1 M phosphate buffer, pH 7.4

(Karlsson and Schultz, 1965).

2.2.15 Electron microscopy

For electron microscopy, fixed slices were washed twice for 10 min with 0.1 M phosphate buffer, incubated for 2 h in 1% OsO_4, followed by three washing steps each for 5 min in 0.1 M phosphate buffer. Slices were then dehydrated by incubating them in increased concentration of ethanol (50%, 70%, 90%, and 100%), each incubated for 10 min. Slices were transferred into glas tubes in order to incubate them in propylene oxyde 2x for 5 min. After stepwise infiltration of the samples with epoxy resin the plastic hardens by heat polymerization. Slices were first infiltrated by stepwise increase in concentration of embedding medium (embedding medium: propylene oxyde 1:2 for 30 min, 1:1 over night , 2:1 for 2 h) before heat polymerization for 24 h at 60 °C. Ultrathin sections were cut using Leica Ultracut S ultramicrotome (Leica, Vienna, Austria) and stained with aqueous 4% uranylacetate followed by lead citrate. The sections were viewed in an electron microscope (Leo EM912AB, Zeiss) and images were taken using on-axis 2048 x 2048 charge coupled device camera (Proscan, Schering, Germany).

3 Results

3.1 MBP accumulates at PIP2 enriched membranes

3.1.1 MBP and PIP2 colocalize at the same subcellular domains

MBP is a protein with a high net positive charge that interacts with acidic phospholipids in model membranes. In cellular membranes, phosphatidylserine (PS) is by far the most abundant negatively charged lipid. Two minor lipids, previously shown to interact with signaling proteins, are phosphatidylinositol(4,5)-bisphosphate (PIP2) and phosphatidylinositol(3,4,5)-trisphosphate (PIP3), which constitute only a few percent of total membrane lipids (McLaughlin et al., 2002). A recombinant pleckstrin homology (PH) domain, derived from one of the two signaling proteins PLCδ1 or AKT, were expressed as a green fluorescent (GFP) and yellow fluorescent protein (YFP) fusion proteins in the oligodendroglial cell-line Oli-neu (Jung et al., 1995), in order to localize PIP2 and PIP3. The PH-domain of PLCδ1 forms a 1:1 complex with PIP2 (Lemmon et al., 1995) and is therefore commonly used as a PIP2 sensor (Milosevic et al., 2005; Rusten and Stenmark, 2006). As shown in Fig. 3.1A, the fusion proteins GFP-PH-PLCδ1 and PH-AKT-YFP, which recognize PIP2 and PIP3, respectively (Várnai and Balla, 1998), were exclusively found at the plasma membrane of oligodendroglial Oli-neu cells. Similar results were obtained when primary oligodendrocytes were induced to exptress GFP-PH-PLCδ1 using the Semliki Forest Virus vector (SFV) (Fig. 3.1B). To study, for comparison, the localization of PS, Lactadherin, with its major PS-binding motif localized to its C2 domain (Lact-C2), fused to a red fluorescent protein (mRFP), was used as a specific PS-sensor (termed mRFP-LactC2; Yeung et al., 2008). In contrast, mRFP-Lact-C2 was targeted to both the plasma membrane and intracellular membranes (Fig. 3.1A). Next, the 14 kDa isoform of MBP was coexpressed together with the fluorescent lipid sensors in Oli-neu cells. Interestingly, MBP staining was most robust at the plasma membrane in regions that also showed an enrichment of PIP2 and PIP3 (Fig. 3.1A). MBP was not associated with intracellular membranes that contained high levels of PS.

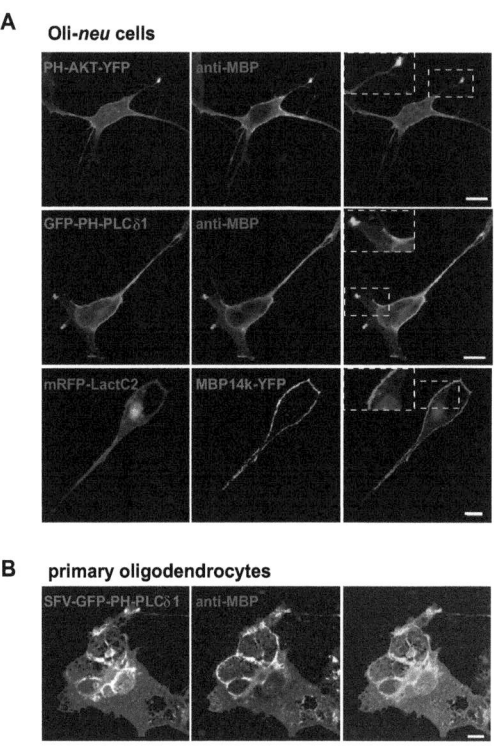

Figure 3.1: MBP colocalizes with PIP2 and PIP3 at the plasma membrane of oligodendrocytes.
(A) Oli-neu cells were transiently cotransfected to express GFP-PH-PLCδ1, PH-AKT-YFP or mRFP-LactC2 to visualize PIP2, PIP3 and PS, respectively with MBP14k (immunolabeled with anti-MBP antibody) or MBP14k-YFP. Colocalization with MBP14k is shown on the right (overlay).
(B) GFP-PH-PLCδ1 was expressed with the SFV vector for 8 h in primary oligodendrocytes. Cells were fixed and immunolabeled against MBP (scale bar, (A) 5 μm, (B) 10 μm.)

This was not due to the physical masking of PS by mRFP-Lact-C2, because MBP showed the same distribution in the absence of any lipid sensor. Similar results were observed for the 18.5 kDa MBP isoform (not shown), while expression of the 21.5 kDa MBP resulted in nuclear staining of some cells as described previously (Pedraza et al., 1997 and data not shown). To localize PIP2 in primary oligodendrocytes, the PH-domain of PLCδ1 fused to GFP was inserted into a SFV-vector. In primary oligodendrocytes, GFP-PH-PLCδ1 did not only localize to the plasma membrane at the cell soma, but was also found in the flat membrane sheets that contain large amounts of MBP, as visualized through immunolabeling MBP (Fig. 3.1B). The preferential association of MBP with membranes of the myelin compartment is achieved, in part, by transport of its mRNA into oligodendroglial cell processes and local translation, which is stimulated by neuronal signals (White et al., 2008). This transport is mediated by a 21-nucleotide RNA transport signal (RTS) in the 3' UTR of the MBP mRNA (Ainger et al., 1997). Importantly, this targeting of MBP to the plasma membrane of Oli-neu cells was independent of the RTS in the 3' UTR, since expression constructs with or without this mRNA targeting signal led to an indistinguishable MBP localization (Fig. 3.1A; MBP without 3' UTR immunolabeled with MBP antibody (red); MBP with 3' UTR was expressed as YFP fusion construct (green) in Oli-neu). Although PS is more abundant at the plasma membrane than PIP2 or PIP3, MBP localized mainly to the plasma membrane at sites of enriched PIP2 or PIP3 levels, independent of the presence or absence of PIP2, PS or PIP3 sensors. Taken together, these data show that MBP does not associate equally well with all membrane surfaces, but has a preference for the plasma membrane, which is enriched in both PIP2 and PIP3, namely the tips of Oli-neu processes and the myelin membrane in oligodendrocytes.

3.1.2 FRET experiments indicate a close association of PIP2 with MBP

FRET (foerster resonance energy transfer) experiments are commonly used to demonstrate the interaction between proteins and have also been widely used to show the interaction between proteins and lipids. FRET is a process, in which the donor molecule is excited and can transfer its excitation energy to the acceptor, if both the donor and acceptor

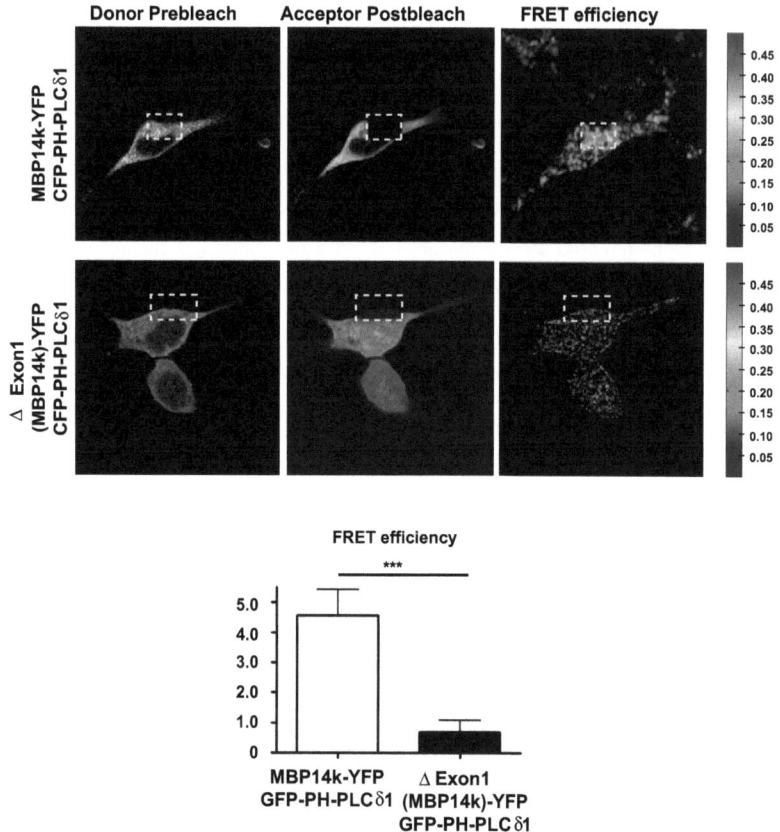

Figure 3.2: FRET imaging revealed an interaction of PIP2 (sensed by CFP-PH-PLCδ1) with MBP, but not with a mutant form of MBP that is unable to bind to the plasma membrane (ΔExon1-MBP14k-YFP; see also Fig. 3.8). Oli-neu cells were transiently transfected to express the indicated plasmids in a 1:1 ratio and fixed after 12 h. FRET was detected by an increase in donor fluorescence after photobleaching of the acceptor. Confocal images of both FRET pairs are shown before (pre) and after (post) photobleaching. FRET efficiency is indicated in pseudocolor (shown from blue to red with increasing FRET efficiency; n > 20 cells; means ± SEM; ***p = 0.0002; t-test).

absorption spectra overlap and if the molecules are in close proximity (between 2-10 nm; Bremer, 2008). The energy transfer is detectable as an increase in fluorescence intensity of the donor. FRET has also been used to measure PIP2 dynamics. CFP-PH-PLCδ1 and YFP-PH-PLCδ1 constructs, co-expressed in the same cell form a FRET pair, since they can be found in the same sub-micrometer domain. Expression constructs of both CFP-PH-PLCδ1 and YFP-PH-PLCδ1 are therefore used to show that PIP2 molecules are clustered in microdomains and to monitor PIP2 dynamics. It is important to note that PH-PLCδ1 interacts with PIP2 even if basic proteins are bound to the latter (McLaughlin et al., 2002; Gambhir et al., 2004). If MBP and PIP2 were to interact with each other by forming clusters, a FRET pair of the PH-domain of PLC and MBP should show energy transfer. I therefore expressed CFP-PH-PLCδ1 and MBP14k-YFP in Oli-neu cells, and indeed energy transfer between both proteins could be detected (Fig. 3.2). The deletion of exon 1 leads to a complete loss of MBP association to the plasma membrane (see experiment below). This truncated form of MBP showed significantly reduced FRET-efficiency compared to full length MBP (Fig. 3.2). This finding confirms that membrane bound MBP associates with PIP2, and that the loss of plasma membrane binding correlates with the loss of MBP-PIP2 interaction.

3.1.3 PIP2 accumulation in endomembranes leads to relocalization of MBP

In order to investigate whether MBP associates to the plasma membrane in a PIP2-dependent manner, a previously published method was used (Brown et al., 2001; Ono et al., 2004): Arf6 is a small GTPase, which is involved in the recycling of early endosomes back to the plasma membrane. Arf6-GTP activates the PIP2 generating enzyme PI4P5K and thereby GTP is hydrolyzed to Arf6-GDP. Since PIP2 is also necessary for endocytosis, Arf6 regulates PIP2 levels at the plasma membrane. In fact, it was shown that overexpression of a constitutive active mutant of Arf6 (Arf6/Q67L) leads to continuous endocytosis together with PI4P5K activity. The overexpression of Arf6/Q67L therefore results in the formation of intracellular vacuoles, which are enriched in PIP2 together with a reduced level of PIP2 at the plasma membrane (Donaldson, 2003). When COS1 cells were cotransfected to express both GFP-PH-PLCδ1 and Arf6/Q67L, the PIP2-sensor (GFP-PH-PLCδ1) mainly localized to large intracellular membrane vacuoles (Fig. 3.3, Brown et al., 2001). In order to investigate, whether overexpression of Arf6/Q67L leads to the relocalization of MBP from the plasma membrane to PIP2 enriched vacuoles, cells were cotransfected to express Arf6/Q67L with a MBP14k-YFP fusion construct. In those cells, MBP accumulated at the rims of vacuoles similar to GFP-PH-PLCδ1 (Fig. 3.3). To illustrate the specificity of PIP2 detection at these vacuoles, a mutant form of GFP-PH-PLCδ1 was expressed, which does not bind to PIP2 due to three point mutations (K30A, K32A, W36N, termed GFP-PH-PLCδ1-3xmut; Milosevic et al., 2005). GFP-PH-PLCδ1-3xmut did not accumulate at vacuoles (Fig. 3.3). Thus, these results suggest that relocalized PIP2 results in a relocalization of MBP to intracellular vacuoles.

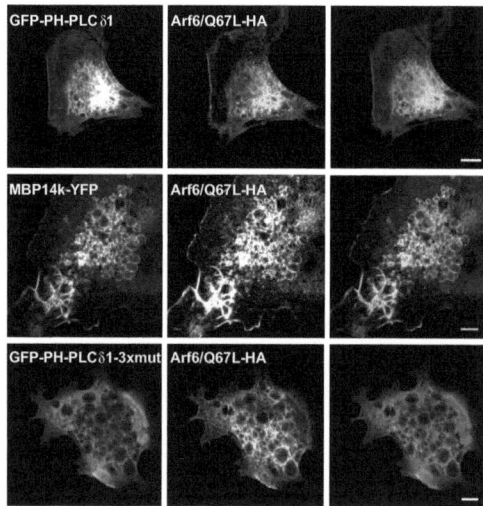

Figure 3.3: Overexpression of constitutive active Arf6 (Arf6/Q67L-HA) in COS1 cells induces accumulation of PIP2-enriched vacuoles and therefore recruits MBP from the plasma membrane to intracellular vacuoles. COS1 cells were transiently cotransfected with plasmids encoding Arf6/Q67L-HA together with either GFP-PH-PLCδ1, MBP14k-YFP or GFP-PH-PLCδ1-3xmut in a 1:1 ratio and fixed after 44 h. Arf6/Q67L-HA was visualized by staining with antibodies against the HA-tag. GFP-PH-PLCδ1 was used to visualize PIP2 in the vacuoles, whereas GFP-PH-PLCδ1-3xmut, which does not bind to PIP2, served as a control (scale bar 5 μm).

3.2 Decreased levels of PIP2 at the plasma membrane leads to decreased MBP binding

3.2.1 Specific hydrolysis of PIP2 leads to reduced plasma membrane association of MBP

In order to test, whether PIP2 and PIP3 levels regulate the plasma membrane association of MBP, PIP2 levels were specifically decreased by expressing synaptojanin 1, one of the major PIP2 hydrolyzing enzymes (mRFP-Synj1, Milosevic et al., 2005) together with MBP14k-YFP, in Oli-neu cells. Additionally, cells were treated with wortmannin (Wm), a PI3K inhibitor to decrease PIP3 levels (Powis et al., 1994). I quantified the amount of MBP at the plasma membrane by measuring the fluorescence intensity of MBP14k-YFP at the plasma membrane in comparison to the cytosol, a method already described for other proteins to quantify their membrane association (Heo et al., 2006). To normalize these results for a known membrane protein, cells were immunostained for the membrane glycoprotein NG2 (AN2-cy5). NG2 staining was then used as a mask to determine the fluorescence intensity at the plasma membrane. The fluorescence intensity at the plasma membrane (PM) was then compared to the intracellular fluorescence intensity (cytosol, cyt) giving a plasma membrane dissociation constant c^{PM}:

$$c^{PM} = \frac{I^{before}(PM)}{I^{before}(cyt)} \frac{I^{after}(cyt)}{I^{after}(PM)}$$

with I^{before} and I^{after} as the fluorescence intensities before and after PIP2 and PIP3 reduction. Although plasma membrane localization of MBP was not completely abolished, the results show a decreased ratio of MBP14k-YFP at the plasma membrane relative to cytosol, compared to control cells that express only MBP (Fig. 3.4A). For comparison, a double-palmitoylated (i.e. membrane-anchored) YFP did not change its membrane localization after coexpression of Synj1 and treatment with Wm (Fig. 3.4A). To test the effect of specific PIP2 hydrolysis on MBP localization in primary oligodendrocytes, these cells were infected with recombinant SFV and expressed a Synj1-GFP fusion construct, fixed and stained against endogenous MBP. Most of the MBP was localized in the cytosol, rather

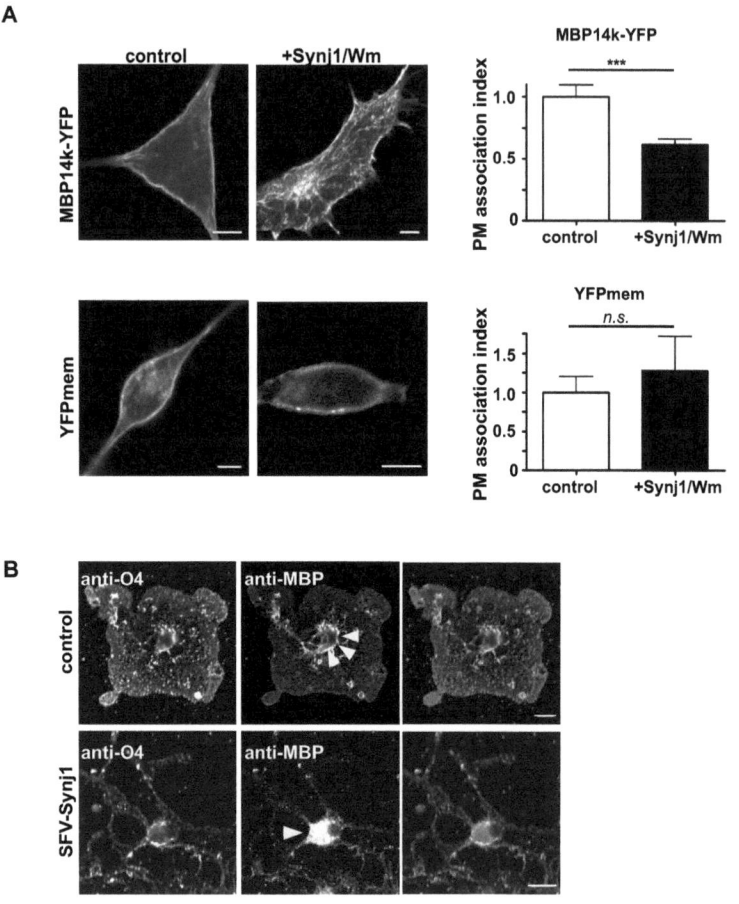

Figure 3.4: PIP2 phosphatase synaptojanin1 (Synj1) reduces MBP localization at the plasma membrane. (A) Oli-neu cells were transiently transfected with Synj1-mRFP and MBP14k-YFP or MBP14k-YFP and vector control in a 1:1 ratio for 12 h before treatment (for 4 h) with the PI3K inhibitor wortmannin (Wm). Cells were fixed and stained against the plasma membrane localized NG2. A plasma membrane association index was calculated as the relative ratio of plasma membrane over cytosolic fluorescence. PIP2 and PIP3 depletion decreased MBP at the plasma membrane, whereas YFPmem localization remained unchanged (n > 70 cells, mean values ± SEM ***p = 0.0006; t-test; scale bar 10 μm).
(B) Primary oligodendrocytes cultured for 4 days in vitro were infected with SFV-Synj1-GFP, fixed after 8 h, and immunolabeled with antibodies against MBP and sulfatide (anti-O4 antibody). Note that expression of Synj1 in oligodendrocytes causes the dissociation of MBP from the cell membrane (indicated by arrowhead) and redistribution into the oligodendroglial soma. O4 immunoreactivity remained unaltered (scale bar 10 μm).

than at the plasma membrane, whereas in control untransfected oligodendrocytes MBP formed a rim demarcating the cell membrane (Fig. 3.4B, arrows indicate rim of cell membrane). Additionally, sulfatide staining, immunolabeled with O4-antibody, was unchanged. These results confirm that the interaction of MBP with the plasma membrane is dependent on the level of PIP2. Since Synj1 specifically hydrolyses PIP2, one can assume that MBP dissociates from the plasma membrane due to reduced PIP2 levels.

3.2.2 PIP2 dependent plasma membrane association of MBP verified in membrane sheets

Membrane sheets are commonly used as a method to visualize events occurring at the intracellular side of the plasma membrane (Lang et al., 2001). Membrane sheets are generated through an ultrasonic pulse. Thereby, the intracellular part of the cell is exposed. Membrane sheets were also used to visualize PIP2 clusters in chromaffin cells (Milosevic et al., 2005). Since MBP in known to associate with the intracellular part of the plasma membrane, membrane sheets were used to confirm the PIP2 dependent association of MBP to the plasma membrane. If MBP associates with the plasma membrane dependent on PIP2 levels, a decreased level of PIP2 should lead to decreased association of MBP at membrane sheets.

When Oli-neu cells were used for the generation of membrane sheets, the application of an ultrasonic pulse led to a detachment of almost all cells from the coverslip and no membrane sheets could be identified. Therefore, another oligodendroglial cell-line (OLN-93) was used for this experiment, since it develops a stronger association to the coverslip than Oli-neu cells (Richter-Landsberg and Heinrich, 1996). In order to decrease PIP2 levels, cells were transfected with the mRFP-Synj1 fusion construct. A decrease in PIP2 level was visualized through coexpression of the PIP2 sensor GFP-PH-PLCδ1. Membrane sheets were generated 16 h after transfection and were identified through the styryl dye TMA-DPH. The association of GFP-PH-PLCδ1 to the plasma membrane sheets was significantly reduced in cells overexpressing Synj1-mRFP (Fig. 3.5). These experiments confirmed that overexpression of Synj1 leads to a decreased PIP2 level at membrane sheets, as previously suggested (Milosevic et al., 2005).

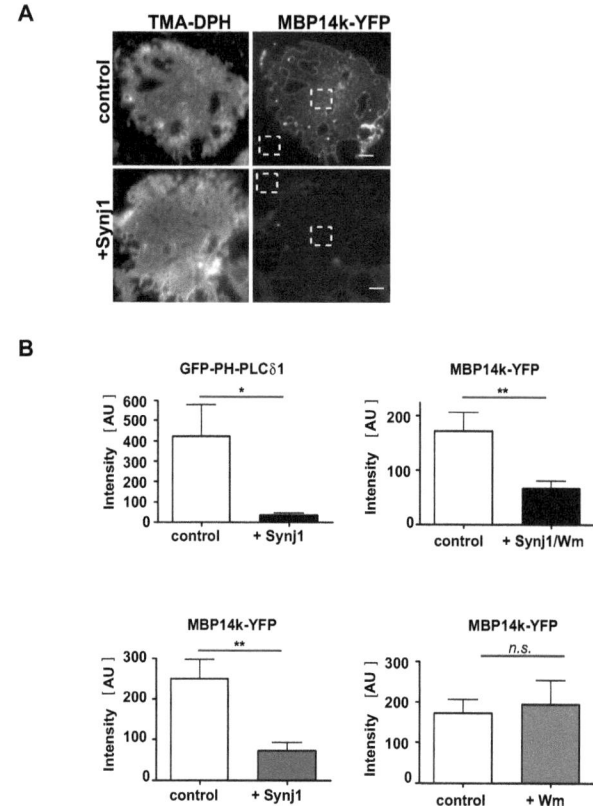

Figure 3.5: PIP2 depletion decreases the association of MBP with plasma membrane sheets. (A)Membrane sheets were generated from transiently transfected OLN-93 cells, expressing MBP14k-YFP and Synj1 or MBP14k-YFP and vector control. Membrane sheets were prepared 16 h after transfection by application of a short ultrasonic pulse. The styryl dye TMA-DPH was used to visualize intact sheets. Images are shown after contrast enhancement. Images showing MBP14k-YFP fluorescence were not contrast enhanced (scale bar 3 μm).
(B) Quantification of MBP binding to the membrane surface relative to background (dotted squares in A). GFP-PH-PLCδ1 was used as a control to indicate the decreased level of PIP2 upon Synj1 coexpression. For PI3K inhibition, cells were treated for 4 h with wortmannin (Wm) 12 h after transfection (n > 100 cells from at least three independent experiments, mean values ± SEM; t-test: GFP-PH-PLCδ1 *p = 0.022; Synj1 expression with PI3K inhibition: **p = 0.003; PI3K inhibition: ns p = 0.747; Synj1 and MBP14k-YFP expression: **p = 0.0019).

3 Results

To test, whether decrease of PIP2 and PIP3 levels also reduces MBP14k-YFP fluorescence, membrane sheets were generated from OLN-93 cells, which overexpressed MBP14k-YFP and mRFP-Synj1. Additionally, these cells were treated with the PI3K inhibitor wortmannin (Wm), which leads to a decrease in PIP3 levels (Powis et al., 1994). Quantification of these experiments demonstrated that a decreased level of PIP2 and PIP3 indeed correlated with a reduced level of MBP at the plasma membrane (Fig. 3.5B).

In order to investigate, whether decreased PIP2 levels are sufficient to reduce the level of MBP associated to the plasma membrane, membrane sheets were generated from cells expressing Synj1-mRFP together with MBP14k-YFP constructs. Overexpression of Synj1 was in fact sufficient to significantly reduce the level of MBP associated to the plasma membrane. This indicates that PIP2 levels alone can control the association of MBP to the plasma membrane.

To test on the other hand, whether a decreased level of PIP3 is also sufficient to displace MBP from the plasma membrane, cells were treated with wortmannin (Wm), to only inhibit PI3K activity (Fig. 3.5). Membrane sheets generated from MBP14k-YFP transfected cells that were treated with Wm, however did not show a significant reduction of MBP from the membrane sheets. However, one cannot rule out a possible involvement of PIP3 in associating MBP to the plasma membrane, since also for other natively unfolded proteins as myristoylated-alanine rich C-kinase substrate (MARCKS) a direct interaction with PIP2 was proposed, but PIP3 also influences the binding to the plasma membrane (McLaughlin et al., 2002; Heo et al., 2006). Since PIP2 is more abundant than PIP3 in quiescent cells, PIP2 is a more efficient binding partner. These experiments verify that MBP associates with the plasma membrane in a PIP2 dependent manner. In all preparations MBP accumulated in clusters similar to those previously observed for PIP2 (Milosevic et al., 2005). It is therefore likely that these clusters are PIP2 enriched domains. Additionally, accumulation of MBP was seen around the edges of the membrane sheets (Fig. 3.5A).

3.3 Decrease in PIP2 at the plasma membrane leads to intracellular accumulation of MBP

In order to investigate, whether MBP associates with the endomembrane system after PIP2 depletion, I stained MBP14k-YFP and Synj1-mRFP cotransfected cells with antibodies against marker-antigens for late-endosome/lysosome (LE/L, Lamp1), Golgi-compartment (BIP) or ER (GM130; Fig. 3.6A). MBP showed no colocalization with any of these markers. This indicates that upon Synj1 expression, MBP forms unspecific clusters within the cytosol. Previously it was shown that the surface charge is regulated by phosphatidyl-serine (PS, Yeung et al., 2008). To visualize PS in cells, the C2 domain of lactatherin fused to mRFP (LactC2-mRFP) was used as PS-sensor, since it specifically binds to PS. When PIP2 levels were decreased at the plasma membrane, MBP colocalized with the PS-sensor (Fig. 3.6B). This finding indicates that PIP2 depletion might lead to binding of MBP with the next available negatively charged lipid, namely PS.

3.4 Replacement of positive amino acids in MBP reduces its binding to the plasma membrane

Since MBP interacts with the plasma membrane electrostatically, alteration of the net charge of MBP should influence this interaction. It would be interesting to identify the region within MBP that interacts with PIP2. I decided to use the 14 kDa isoform for the following experiment, since this isoform bears the least exons of all isoforms (exon 1, 3, 4, 5, 7), but also associates with the plasma membrane and is sufficient to rescue the shiverer phenotype (Kimura et al., 1989). In order to test, which part of the 14 kDa MBP is sufficient for plasma membrane association, I subsequently deleted different exons and expressed these various mutant constructs in Oli-neu cells. A truncated protein, in which exons 1, 3 and 5-encoded regions were deleted, showed no plasma membrane localization (not shown). When exon 1-encoded region alone was deleted (Δ-Exon1(MBP14k)-YFP), this mutant was not able to associate to the plasma membrane (Fig. 3.8) as indicated through line scans across the plasma membrane of 30 different cells. Conversely, when exon 1 encoded region fused

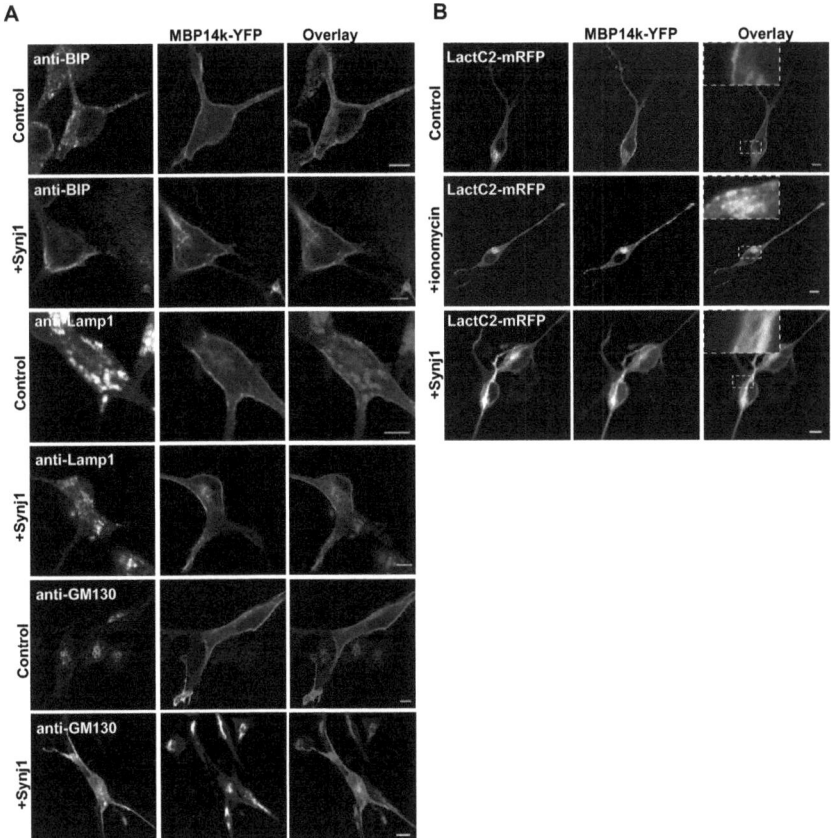

Figure 3.6: Localization of MBP after specific depletion of PIP2 through Synj1 co-expression. (A) Oli-neu cells were cotransfected to express MBP14k-YFP and synaptojanin1 or vector control respectively for 12 h, fixed and immunolabeled against endomembrane markers (BIP - ER, Lamp1 - LE/L, GM130 - golgi; scale bar 5 μm).
(B) PIP2 depletion results in a visible association of MBP to PS containing intracellular membranes. Oli-neu cells were transiently transfected with MBP14k-YFP, LactC2-mRFP and Synj1 and fixed after 16 h. Those cells that were treated for 2 min with ionomycin before fixation (see also Fig. 3.9) also showed MBP localization at phosphatidylserine containing membranes (scale bar 5 μm).

3 Results

to YFP was expressed in Oli-neu cells (termed Exon1(MBP)-YFP), this mutant associated to the plasma membrane, although to a less extent than the full-length encoded MBP14k-YFP construct (Fig. 3.8). For comparison, the exon 7-encoded region was not sufficient to bind to the plasma membrane (termed Exon7(MBP)-YFP). These data suggest that the N-terminal domain (encoded in exon 1) is sufficient to bind to the plasma membrane. However, we found a strong reduction of MBP at the plasma membrane compared to full-length 14 kDa MBP, which indicates that a reduction of the net positive charge of MBP leads to a reduced membrane binding. When the amino acid sequences of the N-terminal domain (encoded by exon 1) were aligned from different species, I noticed that many positively charged residues were strictly conserved (Fig. 3.7). To test for the functional significance of these positions in transfected cells, MBP mutants were generated in which one or two (closely spaced) basic amino acids were replaced by alanine (Fig. 3.8).

Unexpectedly, in the majority of cases (K5A/R6A; H22A/R24A; R30A/R32A; R42A;

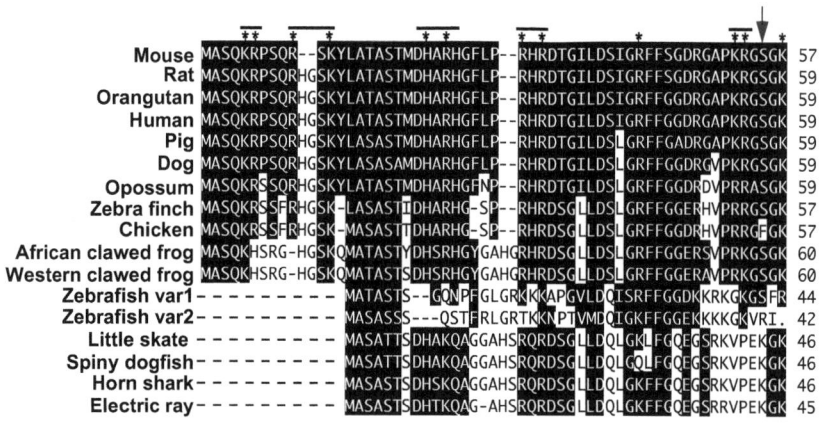

Figure 3.7: MBP sequence alignment of the N-terminal domain (encoded by exon1) with selected orthologs. Sequences were retained from NCBI database and clustal-alignment with PAM250. Amino acids identical to mouse MBP are in black boxes. Basic amino acids analyzed by site-directed mutagenesis are marked by asterisk, a serine residue altered by mutagenesis is marked with an arrow (this figure was composed by HB Werner, Dpt Neurogenetics, MPI-em, Göttingen).

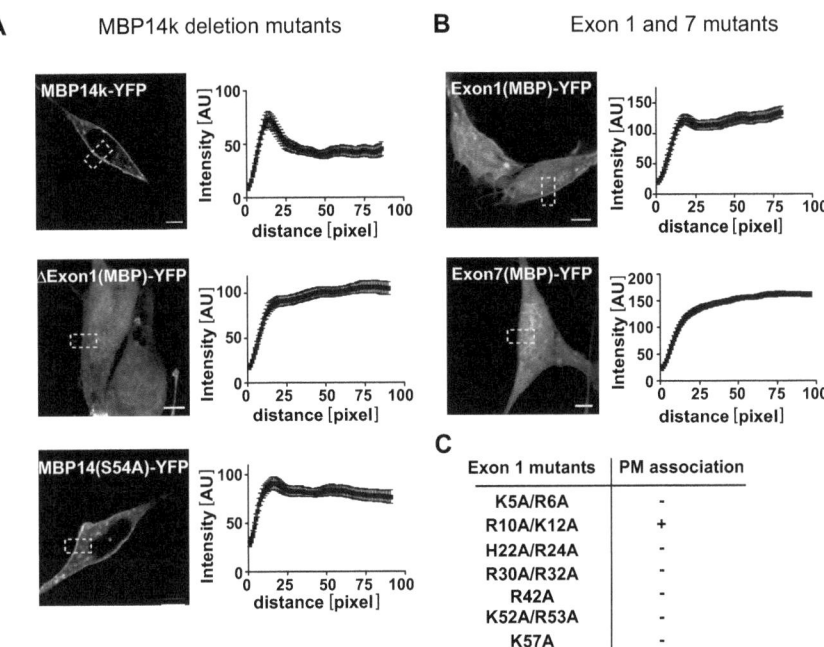

Figure 3.8: Quantification of membrane localization of truncated MBP isoform. (A, B) Shown are cells expressing the deletion construct lacking the exon 1-encoded region (ΔExon1(MBP)-YFP), or expressing the exon 1-encoded region fused to YFP (Exon1(MBP)-YFP) or expressing the exon 7-encoded region fused to YFP (Exon7(MBP)-YFP). S54A mutant shows less plasma membrane association than wild type MBP14k-YFP (scale bar 5 μm). Various positive amino acids (Arg, His or Lys) were replaced for Ala. Intensity profiles were generated for each cell. Presented are the mean intensity profiles from 30 different cells with standard deviations. (C) The table indicates which mutant protein associates with the plasma membrane.

K52A/R53A; K8A) the replacement of only one or two positively charged amino acids in the N-terminal domain of MBP was sufficient to prevent protein binding to the membrane, while the exchange of two other basic residues within exon 1 (R10A/K12A) did not effect plasma membrane localization (Fig. 3.8). Previous biochemical studies had suggested that PIP2 is covalently attached to S54 (Chang et al., 1986; Yang et al., 1986). I therefore generated a mutant, in which I replaced S54 for alanine within full-length MBP14k, as well as within the truncated exon 1-encoded region. These S54A-mutants showed a reduced binding compared to wild-type construct. This indicates that S54 influences the binding capacity to the plasma membrane, however it is not sufficient to completely abolish the ability of MBP to bind to the plasma membrane (Fig. 3.7 arrow, Fig. 3.8 and not shown).

Taken together, the alteration of the net positive charge of MBP influences the binding to the plasma membrane. This implies that MBP interacts with PIP2 electrostatically, as previously suggested (Harauz et al., 2004). Additionally, the localization of basic residues seems to be important for the binding of MBP to PIP2. The tertiary structure might play a crucial role for the natively unfolded protein MBP. One putative binding domain was found within exon-1 encoded region.

3.5 Membrane surface charge influences the plasma membrane localization of MBP

The electrostatic attraction of basic proteins to the negatively charged membrane is dependent on the surface charge of the plasma membrane. Since PIP2 has a valence of -4 at physiological pH, decrease in negatively charged lipids such as PIP2, leads to a decrease in surface charge (Yeung et al., 2008). It has become apparent that the surface potential is influenced by PIP2 due to its high valence, compared to other negatively charged lipids that have a valence of -1. The plasma membrane in comparison to endomembranes is therefore more negatively charged, due to accumulation of the negatively charged lipid PIP2. In order to characterize the interaction of MBP with the plasma membrane in more detail, the surface charge of the plasma membrane was altered, as previously described (Yeung et al., 2006). It was shown that the PIP2-sensor GFP-PH-PLCδ1 dissociates from the plasma membrane upon Ca^{2+} influx and endogenous PLC activation (Várnai and Balla, 1998). PLC activity in turn, leads to PIP2 hydrolysis and flipping of PS to the extracellular part of the membrane. Both processes lead to an overall decrease in surface charge at the plasma membrane (Yeung et al., 2006b; Bevers et al., 1983).

Ionomycin is a highly selective Ca^{2+} ionophore (Liu and Hermann, 1978). The increase of intracellular Ca^{2+} concentration upon ionomycin treatment leads to PLC activation and subsequent PIP2 hydrolysis. This method had previously been used to describe the specificity of the PIP2-sensor GFP-PH-PLCδ1 (Várnai and Balla, 1998). Oli-neu cells were transfected

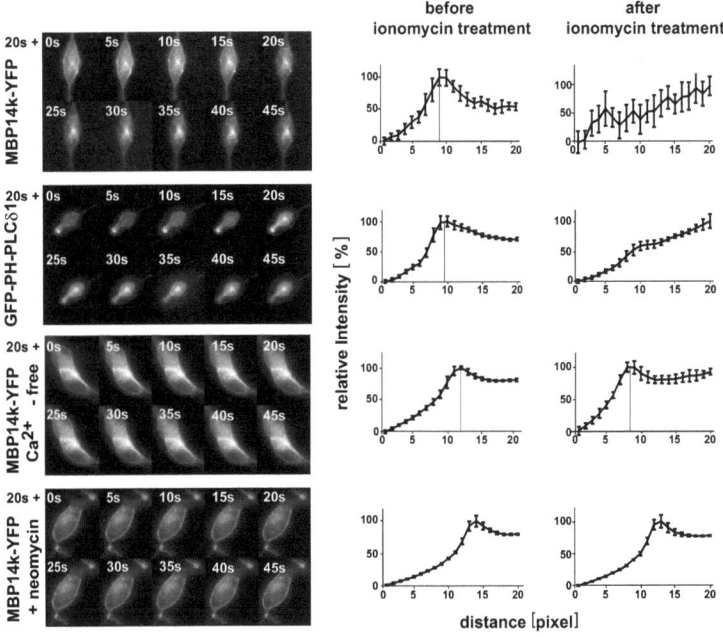

Figure 3.9: Decrease in surface charge displaces MBP14k-YFP from the plasma membrane of Oli-neu cells. Oli-neu cells were transfected with MBP14k-YFP or GFP-PH-PLCδ1 and subjected to live cell imaging. Cells were bathed in medium with or without Ca^{2+}. 20 s after addition of 10 μM ionomycin images were obtained every 10 s. To block PLC dependent PIP2 hydrolysis, neomycin (10 mM) was added to the culture medium 10 min before acquisition of images and ionomycin treatment. Shown are the line scans of cells before and after ionomycin treatment.

with the GFP-PH-PLCδ1 fusion construct and imaged in medium containing 2 mM Ca^{2+} as described before (Várnai and Balla, 1998). As a result of PIP2 hydrolysis, addition of ionomycin to the imaging solution induced a rapid dissociation of GFP-PH-PLCδ1 from the plasma membrane (Fig. 3.9) within 60 sec. When cells were bathed in Ca^{2+}-free medium, the PIP2 sensor remained at the plasma membrane even after 15 min of ionomycin application. In order to test, whether PIP2 hydrolysis induces dissociation of MBP from the plasma membrane, cells were transfected with MBP14k-YFP construct and treated with ionomycin. Application of ionomycin to the imaging solution induced a rapid dissociation of MBP from the plasma membrane (Fig. 3.9). Like the PIP2 sensor, MBP also dissociated from the plasma membrane within 60 sec. These experiments were quantified by plotting the fluorescence intensity across the plasma membrane. Shown are the mean intensity plots from 30 cells with respective standard deviations (Fig. 3.10). In order to investigate, whether intracellular pools of Ca^{2+} are sufficient to dissociate MBP from the plasma membrane, the same experiment was performed in cells that were kept in modified Krebs-Ringer solution without Ca^{2+}. These experiments showed that ionomycin treatment had no effect on plasma membrane localization even after 30 min of incubation, which illustrates that MBP only dissociates from the plasma membrane as result of influx of Ca^{2+}. Intracellular pools were not sufficient to induce this effect (Fig. 3.9). Neomycin forms a 1:1 electroneutral complex with PIP2 (Gabev et al., 1989; Arbuzova et al., 2000) and thereby blocks PLC dependent PIP2 hydrolysis (Várnai and Balla, 1998). The PIP2 sensor therefore did not dissociate from the plasma membrane in cells that were pre-incubated with neomycin before ionomycin treatment. In order to test, whether MBP dissociated from the plasma membrane due to PLC activation and subsequent PIP2 hydrolysis, cells were pre-incubated with neomycin before ionomycin treatment. When PIP2 was blocked through neomycin, MBP remained at the plasma membrane. Pre-incubation of cells with neomycin therefore completely prevented Ca^{2+} induced dissociation of MBP from the plasma membrane. Additionally, if cells were treated for 30 min with EGTA after 2 min exposure to ionomycin, MBP relocalized to the plasma membrane. Although EGTA is not plasma membrane permeable, cells were able to pump excess Ca^{2+} out of the cell, which led to the re-association of both GFP-PH-PLCδ1 or MBP14k-YFP, since addition of EGTA was sufficient to chelate excess Ca^{2+} (Fig. 3.10). When EGTA was added to the cells in a Ca^{2+}-free buffer, no effect on the localization of

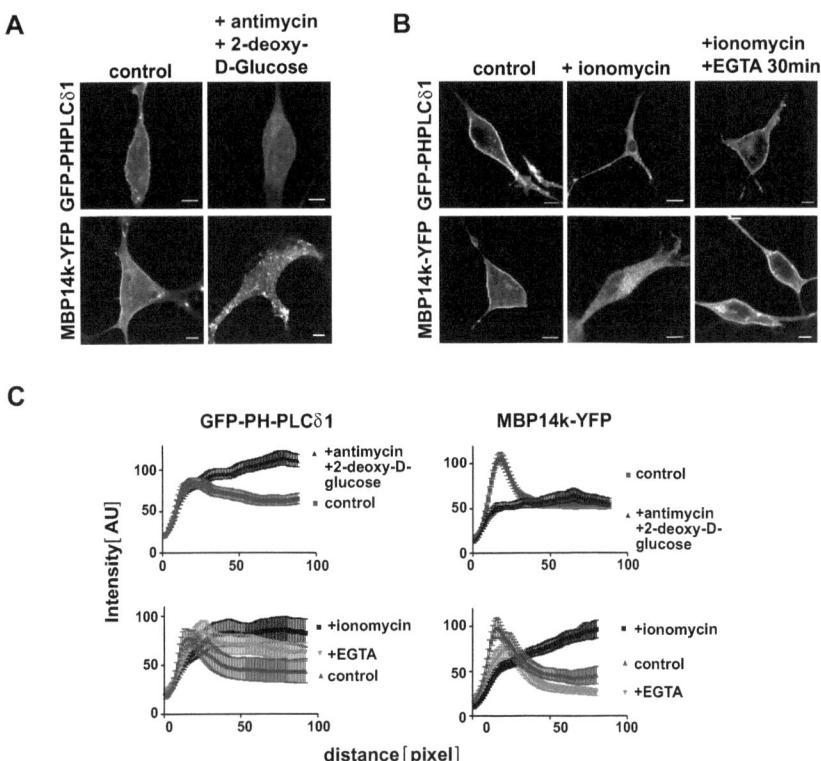

Figure 3.10: (A, B) Surface charge reduction of MBP14k-YFP or GFP-PH-PLCδ1-transfected Oli-neu cells through treatment with antimycin and 2-deoxy-D-glucose (which prevents new synthesis of PIP2) or ionomycin (which increases intracellular Ca^{2+}; scale bar 5 μm).
(C) MBP localization was quantified from confocal images taken from fixed and mounted Oli-neu cells that were treated with ionomycin, EGTA after ionomycin treatment, antimycin/2-deoxy-D-glucose or control medium respectively. Shown are line scans over the plasma membrane of at least 10 individual cells with respective standard deviations (n > 10).

MBP or GFP-PH-PLCδ1 was observed (data not shown). I therefore conclude that MBP dissociates from the plasma membrane, due to Ca^{2+} influx and subsequent PLC activation. I also found similar results for exon 1 encoded region (Exon1(MBP)-YFP; data not shown).

In order to test whether the alteration of surface charge releases MBP from myelin sheets of primary oligodendrocytes, cells were first treated with ionomycin for 2 min and then permeabilized with saponin, to wash out released protein. Cells were then fixed and immunolabeled against GalC (anti-O1) and MBP (Fig. 3.11). Indeed, MBP was almost completely washed from the myelin sheets, whereas O1 staining remained unaltered compared to control DMSO-treated cells.

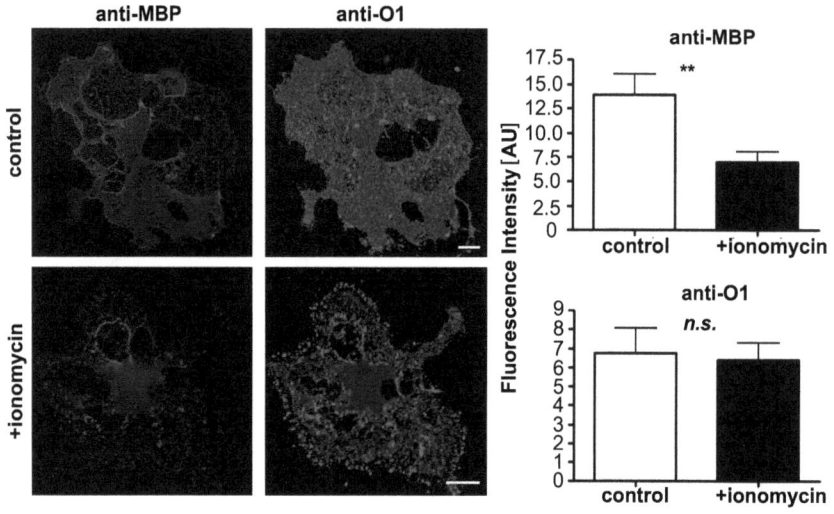

Figure 3.11: Release of MBP from myelin membrane upon Ca^{2+} treatment. Primary oligodendrocytes that were treated with ionomycin or not, were permeabilized with saponin to wash out released proteins, fixed and immunolabeled against MBP and GalC (O1). Fluorescence intensity of MBP-cy3 and O1-cy2 was quantified relative to background (n = 30, from two independent experiments; p = 0.0039; scale bar 5 μm).

A second method to reduce PIP2 levels and thereby change the surface charge at the plasma membrane is to deplete cells from ATP. Depletion of ATP leads to decrease in phosphoinositide levels, since the phosphorylation of the ionositol ring of PI is ATP-dependent. ATP depletion was shown to lead to a decrease in PIP2 after 45 min (Yeung et al., 2006b). Cellular ATP is decreased by culturing cells in Glucose and Ca^{2+}-free medium together with antimycin (which blocks complex III of the mitochondrial electron transfer chain) and 2-deoxy-D-glucose (which cannot be processed in the glycolysis cycle). In order to test, whether ATP depletion of Oli-neu cells leads to a decrease in PIP2 levels, I transfected cells again with GFP-PH-PLCδ1 fusion construct, to visualize PIP2. Cells were then cultured in ATP depletion medium. ATP depletion reduced the plasma membrane-association of the PIP2-sensor (GFP-PH-PLCδ1) as indicated in the line scan plot taken from >10 different cells over their plasma membrane, as previously described (Fig. 3.10C; Yeung et al., 2006). In order to test, whether ATP depletion also results in loss of MBP association to the plasma membrane, Oli-neu cells were induced to express MBP14k-YFP fusion construct. ATP depletion of these cells resulted in a reduced binding of MBP to the membrane. However, Ca^{2+} and glucose-free medium, without the addition of antimycin and 2-deoxy-D-glucose was sufficient in some cells to lower the association of MBP to the plasma membrane, indicating that MBP might be highly sensitive to any changes in available ATP.

Thus, both ATP depletion and ionomycin experiments illustrated that MBP depends on the negative charge present on the intracellular surface of the plasma membrane, which in turn is influenced by PIP2 levels.

3.6 Role of MBP-PIP2 binding for the maintenance of myelin integrity

The decrease in negative surface potential (Fig. 3.10) showed that the binding of MBP to the plasma membrane is dependent on the negative charges at the intracellular membrane surface in a PIP2-dependent manner. In order to investigate the role of surface charge in myelin, acute slices from corpus callosum were treated with either ionomycin or depleted of ATP (Fig. 3.12). We observed a vesiculation of myelin membranes and loss of compaction in these slices (Fig. 3.12 and Fig. 3.13). Previous studies had shown altered neurofilament spacing in oxygen-glucose deprived slices and as a result of energy deprivation, similar vesiculation was observed (Tekkök et al., 2007). Conversely, when slices were pre-incubated

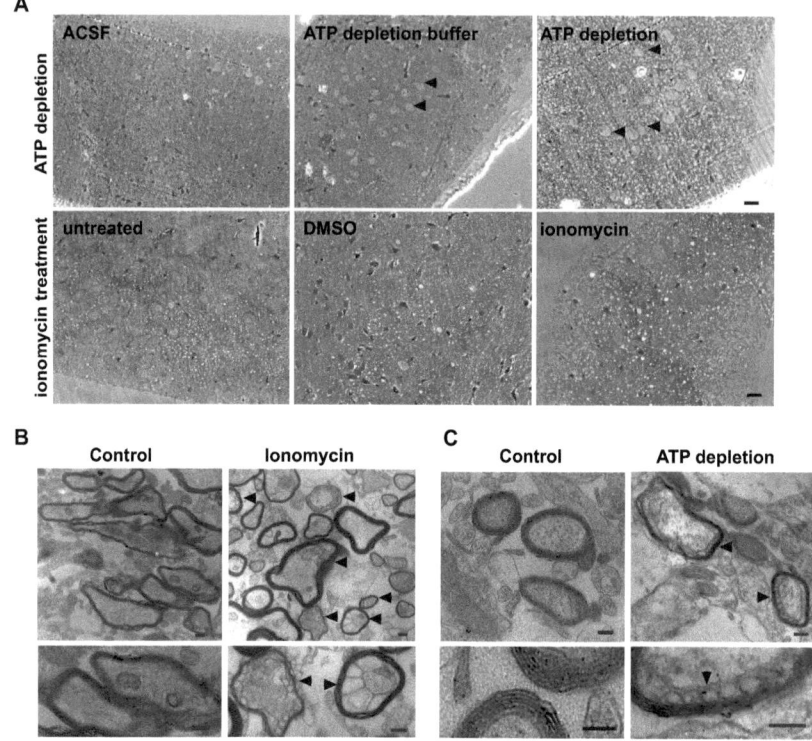

Figure 3.12: Vesiculation of myelin in acute brain slices following ionomycin treatment or ATP depletion. Acute slices from 30 days old mice were equilibrated in ACSF for 1h at room temperature before incubated with ionomycin or ATP depleting reagents for 1h at 35 °C. For ionomycin treatment control slices were incubated in DMSO. For ATP depletion acute slices from adult mice were incubated in glucose and Ca^{2+} free ACSF together with antimycin and 2-deoxy-D-glucose for 1h. Control slices were incubated in ACSF. After 1h incubation, acute slices were fixed and processed for electron microscopy.
(A) Semi-thin sections were labeled with lyxol fast blue and imaged with a Leica epifluorescent microscope. Arrows indicate morphological differences between treated and untreated slices (scale bar 20 μm).
(B, C) Electron micrograph of acute slices. Arrowheads indicate myelin delamination (scale bar (B) 0.6 μm, (C) 0.4 μm).

with neomycin, to prevent PLC-mediated PIP2 hydrolysis, the number of vesiculated myelin profiles segments was significantly reduced (Fig. 3.13).

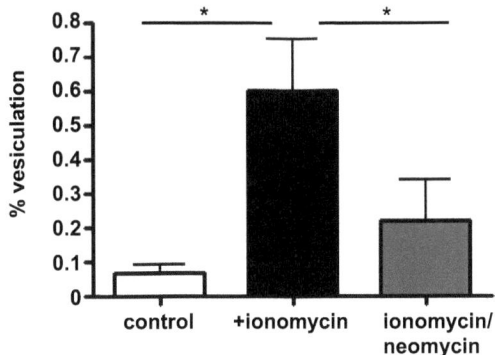

Figure 3.13: Vesiculation was less pronounced when slices were incubated for 15 min with 10 mM neomycin prior and during ionomycin treatment (for 30 min, control vs. ionomycin: p = 0.00129; control vs. ionomycin/neomycin: p = 0.0338; n >100 myelinated axons; 4 independent slices).

3.7 A hypothetical role of MBP in regulating membrane tension

PIP2 not only plays a crucial role in determining surface charge. It was shown that PIP2 also regulates the adhesion of the cytoskeleton to the plasma membrane (Raucher et al., 2000). PIP2 regulates the membrane tension by anchoring actin-binding proteins to the plasma membrane. The reduction of the available pool of PIP2 through overexpression of the PH-domain of PLCδ1 therefore prevents actin binding proteins from binding to PIP2 and leads to a reduction of adhesion energy, as measured by atomic force microscopy (Raucher et al., 2000). A sign of alteration of membrane-cytoskeleton adhesion is an increase in membrane bleb formation. If we suppose that for the formation of myelin, membrane tension has to be reduced, MBP interaction with PIP2 might have a regulatory role in plasma membrane tension and membrane growth (Czech, 2000).

3 Results

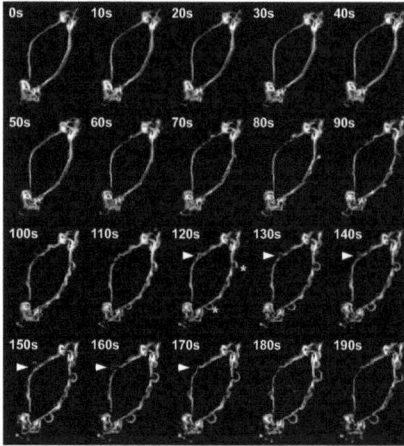

Figure 3.14: Oli-neu cells induced to express MBP14k-YFP for 8 h and subjected to ionomycin treatment. 20 sec after application of ionomycin to the culture medium, images were obtainied every 10 sec. During life cell imaging, membrane bleb formation was oberved (asterisk). Additionally, prior to bleb formation a sudden increase in small protrusions was found (arrow), which were partially retracted over time (Images were modified with an ImageJ plugin to visualize the cell's boundary).

In order to test this hypothesis I expressed the MBP14k-YFP fusion construct in Oli-neu cells and treated the cells with Latrunculin A, which depolymerizes F-actin and was shown to lead to bleb formation. Latrunculin A treatment had no effect on the localization of MBP, which indicates that MBP binds to the plasma membrane in an actin independent manner (data not shown). A frequent sign of reduction in membrane tension is the formation of membrane blebs. In Oli-neu cells that express MBP14k-YFP, ionomycin treatment led to a sudden increase of small protrusions (Fig. 3.14). This process was followed by bleb formation.

3.8 Characterization of oligodendrocytes during myelination *in vitro*

Different in vitro culture systems have been developed, in which oligodendrocytes can myelinate axons (Lubetzki et al., 1993; Kleitman et al., 1998; Trajkovic et al., 2006; Chan et al., 2006; Taveggia et al., 2008). I used such a culture system to study the process of myelination in more detail. Cells derived from mouse brain homogenate were plated onto poly-L lysine coated cover-slips and allowed to differentiate until the neuronal network was completely established. Oligodendrocyte precursor cells were then plated on top of the established neuronal network and allowed to differentiate for five days (Trajkovic et al., 2006). During this period, oligodendrocytes extend their processes towards the axons and form the myelin membrane (Fig. 3.15). The different stages of oligodendrocytes can be distinguished through different stage-specific markers. Co-cultures were fixed and stained at different times during the myelination process (8 h, 24 h, 48 h and 5 days). 8 h and 24 h after plating oligodendrocyte precursor cells onto the neuronal network, early progenitor oligodendrocytes were labeled against NG2. At 48 h cells were identified through the O4 antibody. The initial contact to the axon could be observed 8 h after plating oligodendrocytes onto the differentiated neuronal network (Fig. 3.15). After 24 h more oligodendrocyte-neuronal contacts had been established and 48 h after plating these initial contacts were formed into membrane extensions surrounding the axons. Although different oligodendrocytes within these cultures differed in their course of development, most oligodendrocytes had enwrapped neurons within five days and were labeled with the MBP antibody. Taken together, in this culture system oligodendrocytes form the initial contact about 8 h after plating and enwrap axons between day two and three in vitro. After five days axons are myelinated. This culture system was modified in the following experiments but was also used for studies of myelination (Brinkmann et al., 2008).

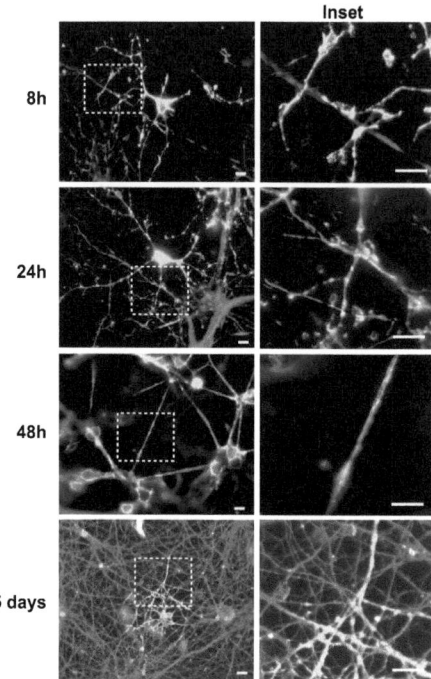

Figure 3.15: Co-cultures can be used as a model system to analyze molecular and morphological processes during myelination. To characterize this culture system in more detail, cells were fixed at different times and immunolabeled against respective marker antigens. Already 8 h after plating oligodendrocytes onto neuronal networks (stained against β-tubulin), a contact to the axon was established (immunolabeled against NG2; at 24 h and 48 h, cells were stained with O4 antibody). In this culture system, oligodendrocytes formed a myelin sheath around axons within 5 days in vitro (immunolabeled against MBP). For different assays, this culture system was modified accordingly and used as a method to study myelination in more detail (see also Brinkmann et al., 2008; scale bar 10 μm).

3.9 Polarization of oligodendrocytes

3.9.1 PIP3 accumulates in Oli-neu cells at the tips of processes

The myelin membrane formed by oligodendrocytes differs from the normal plasma membrane in its lipid as well as its protein composition. Therefore, directed transport of proteins as well as lipids toward the myelin membrane is a prerequisite for proper myelin formation. The specialization of myelin is thus commonly compared to the specification of axons (Simons and Trotter, 2007; Maier et al., 2008). Additionally, similar to neurons, oligodendrocytes form processes that are later retracted, which indicates that the membrane extensions that later form the myelin sheath might also accumulate polarization factors. To investigate the distribution of PIP3 and their downstream proteins in oligodendrocytes, the oligodendroglial precursor cell line Oli-neu was used. Treatment of Oli-neu cells with medium from

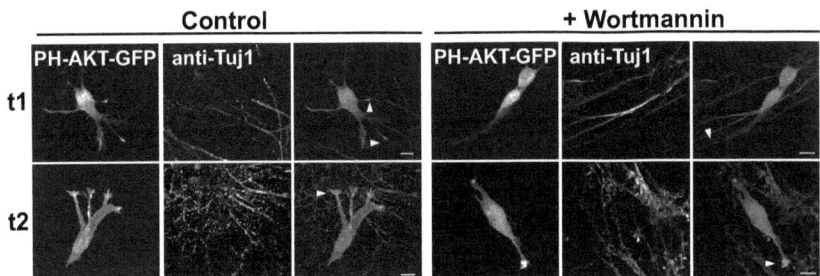

Figure 3.16: A Modified co-culture system as shown in Fig. 3.15 was used for this experiment. The PIP3 sensor PH-AKT-YFP was used to visualize the distribution of PIP3 in the oligodendrocyte precursor cell line Oli-neu, when these cells were co-cultured with neurons. Oli-neu cells were transfected with PH-AKT-YFP and plated onto neuronal cultures. PIP3 mainly accumulated at the tips of processes and at sites of contact with axons (labeled for β-tubulin, Tuj1). Upon wortmannin (Wm) treatment, PIP3 accumulation could not be detected at the tips (after t1=4 h, 30 nM). To test, whether Wm treatment resulted in cell death, Wm was washed out of the culture medium (t2=12 h) and the PIP3-sensor reappeared at the tips of processes. The cells still displayed a normal morphology (scale bar 5 μm).

neuronal cultures results in an increase in number of processes and a redistribution of myelin proteins (Trajkovic et al., 2006; Kippert et al., 2007). When plated on top of differentiated neuronal cultures, Oli-neu cells formed multiple processes (Fig. 3.16). When Oli-neu cells were transfected to express the PIP3-sensor (PH-AKT-YFP) and seeded on top of the neuronal network, PIP3 was found accumulated mainly at the tips of the processes in regions of contact with axons. In order to test the specificity of the PIP3 sensor, Oli-neu/neuronal co-cultures were treated with the PI3K-inhibitor wortmannin (Wm). Wm-treatment completely abolished the accumulation of the PIP3 sensor at the tips, suggesting that PH-AKT specifically recognizes PIP3 in our system.

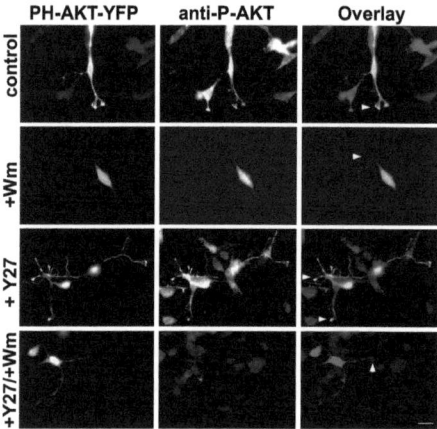

Figure 3.17: Due to PI3K activity, PIP3 is mainly found at the tips of Oli-neu processes, as visualized with the PIP3 sensor PH-AKT-YFP. Additionally downstream of PI3K, AKT is phosphorylated (P-AKT) and co-localizes with PIP3. Inhibition of PI3K through wortmannin (Wm), reduced PH-AKT accumulation, indicating that the PIP3 sensor specifically binds to PIP3. ROCK inhibition (Y27) leads to an increase in the number of processes. PIP3 accumulated at the tips of the newly formed processes, but was again lost upon Wm treatment (scale bar 5 μm).

In other cell types, PI3K inhibition through Wm treatment (at concentrations above 50 nM) was shown to induce apoptosis (Vemuri et al., 1996; Padmore et al., 1996). In order to investigate, whether the loss of PIP3 at the tips was due to induction of cell death, Oli-neu cells were first treated with Wm (the PI3K inhibitor, time point t1=4h) and Wm was then subsequently washed out of the culture medium (time point t2=12h). After Wm was washed out, PH-AKT-YFP reappeared at the tips of Oli-neu cells, which indicates that Wm treatment did not result in induction of apoptosis and that PH-AKT-YFP accumulation at the tips of processes correlates with PIP3 accumulation.

3.9.2 Rho inhibition correlates with PIP3 accumulation at the tips of cellular processes

Inhibition of ROCK induces the formation of processes in Oli-neu cells and results in a redistribution of myelin proteins, similar to a treatment with neuronal conditioned medium (Kippert et al., 2007). In order to investigate, whether ROCK inhibition also leads to accumulation of PIP3 at tips of Oli-neu processes, ROCK was inhibited in PH-AKT-YFP transfected Oli-neu cells. PIP3 was found accumulated in most tips of Oli-neu processes, independent on the number of membrane extensions (Fig. 3.17). When cells were treated with both ROCK and PI3K inhibitor, an increase in the number of processes was observed (as a result of ROCK inhibition), but these processes did not accumulate PIP3 (due to PI3K inhibition). It was shown, that in hippocampal cultures, PIP3 accumulation leads to the targeting of polarization factors, such as the mPar3/mPar6 complex and P-AKT, at the tips of the axonal growth cone (Shi et al., 2003; Ménager et al., 2004). In order to test, whether downstream molecules of PI3K are also accumulated in oligodendrocytes, Oli-neu cells were treated with ROCK inhibitor and stained against phospho-AKT. The PIP3-sensor PH-AKT-YFP colocalized with phospho-AKT (Fig. 3.17). Accumulation of Par3 and phospho-AKT was also found in cells treated with neuronal conditioned medium (Fig. 3.18). However, no difference in staining between cells cultured in normal growth medium and cells cultured in neuronal conditioned medium were observed.

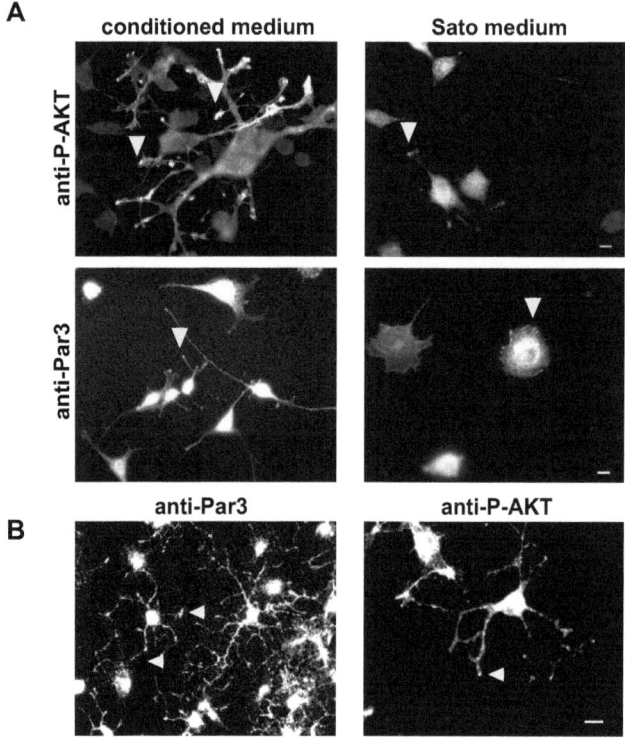

Figure 3.18: Downstream of PI3K activity, polarizing factors, such as Par3 and P-AKT are accumulated at the tips of processes. (A) Oli-neu cells were either incubated in neuronal conditioned medium, which was shown to induce an increase in the number of processes and drive differentiation of Oli-neu cells, or with conventional culture medium. In both cases downstream of PIP3 accumulation phosphorylated AKT (P-AKT) as well as Par 3 accumulated at tips of Oli-neu processes.
(B) Primary oligodendrocytes were cultured for 2 days, fixed and immunolabeled for polarizing factors that were shown in other cell types to accumulate at the tips of processes as a result of polarization. Indeed, also in oligodendrocytes, Par3 and P-AKT accumulated at the tips of process (scale bar 10 μm).

In order to test, whether Par3 or phopho-AKT accumulate at tips of processes prior to myelin membrane formation, I cultured primary oligodendrocytes for 2 days in vitro and stained them against phopho-AKT and Par3. I chose this time, because oligodendrocytes had not established a myelin membrane but had formed process extensions. Par3 and phospho-AKT in primary oligodendrocytes cultured for two days in vitro were mostly found at the tips of processes (Fig. 3.18B). Taken together these experiments provide first insight into polarization signals that might occur during the development of oligodendrocytes.

4 Discussion

4.1 PIP2 dependent association of MBP to the plasma membrane

Although MBP has been investigated for many years, mechanistic insight into MBP function has not been reached, except for its rather unspecific interaction with negatively charged lipid head groups. While mutant mice lacking MBP expression (shiverer) have revealed a protein function in myelin compaction, these in vivo experiments failed to provide mechanistic insight at the molecular level. I have therefore turned to an intact cellular system, in which the presumed interaction of MBP with phospholipids can be studied. Specifically, I used different experimental approaches to investigate the role of PIP2 for the stable association of MBP with the plasma membrane. In a loss-of-function approach, I found that coexpression of MBP with the major PIP2 hydrolyzing enzyme, Synj1, reduced the binding of MBP to the plasma membrane. Additionally, coexpression of MBP with a constitutive active Arf6 variant resulted in the redistribution of MBP from the plasma membrane to intracellular PIP2-enriched endosomal vesicles. Furthermore, I observed FRET between MBP and PIP2-sensing probe, indicating a tight colocalization. A putative PIP2 binding domain of MBP was localized at the N-terminal domain, encoded by exon 1 (contained in all MBP splice isoforms). This region harbors the critical lysine and serine residues. Finally, I found that increasing the intracellular Ca^{2+} level causes a rapid dissociation of MBP from the plasma membrane, which involves the PLC-dependent hydrolysis of PIP2.

Taken together, these results provide experimental evidence that PIP2 is the critical membrane lipid required for the membrane association of MBP. Abnormal Ca^{2+} entry and Ca^{2+}-dependent myelin delamination in white matter tracts, is thus likely to be caused by the detachment of MBP from myelin membranes, which becomes clinically relevant after hypoxic injury or an autoimmune attack.

4.2 Possible roles for MBP-PIP2 interaction

4.2.1 PIP2 as a targeting-signal to the plasma membrane

PI4P and PIP2 are the major phosphoinositides in cells, constituting 1% of total lipid (Gamper and Shapiro, 2007a). Previous work had suggested that proteins with polybasic clusters are targeted to the plasma membrane through interaction with PIP2 and PIP3 (Heo et al., 2006). Since PI(4)P, PI(4,5)P2 and PI(3,4,5)P3 are mainly found at the plasma membrane, these lipids are thought to direct specific proteins to the cell surface (Czech, 2000; Krauss and Haucke, 2007). Recent publications have suggested that PIP2 discriminates endomembranes from the plasma membrane (Yeung et al., 2008). PIP2 was shown to cluster into microdomains when bound to basic proteins (Gambhir et al., 2004; Wang et al., 2004; Golebiewska et al., 2006) and has been suggested to accumulate at the myelin membrane (Baumann and Pham-Dinh, 2001). Therefore it seemed as a possible interaction partner of MBP. Here, I was able to show that MBP binding to the plasma membrane is dependent on the presence of PIP2: expression of a constitutive active mutant of the small GTPase Arf6 (Arf6/Q67L), a regulator of PIP2 levels at the plasma membrane, induced accumulation of PIP2 in endomembranes (Donaldson, 2003; Ono et al., 2004). As a result of PIP2 accumulation, MBP localized to these endomembranes. These findings indicate that MBP associates to the plasma membrane most likely due to its unique composition in lipids. Since MBP is a highly basic protein, it is extremely adhesive to negatively charged lipids (Smith, 1992; Rivas and Castro, 2002; Haas et al., 2007; Rispoli et al., 2007). It therefore seems apparent that translation of MBP is restricted to its final location (Barbarese et al., 1999; White et al., 2008), since MBP could otherwise accumulate in the intracellular membrane system or interact with nuclear proteins. Indeed, in our study, specific hydrolysis of PIP2 through overexpression of the PIP2 phosphatase Synj1, caused MBP to associate with endomembranes that are enriched in PS.

4 Discussion

4.2.2 The Role of MBP in organizing myelin lipids into microdomains

PIP2 is thought to be an important signaling molecule, since it plays a role in the attachment of the cytoskeleton, exo- and endocytosis, membrane trafficking, phagocytosis, protein targeting, and activation of enzymes (Fig. 1.3; Czech, 2000; McLaughlin et al., 2002). It was also shown to induce a change in conformation of ion channels (Hilgemann et al., 2001; Runnels et al., 2002; Prescott and Julius, 2003; Rapedius et al., 2007). In recent years it has become apparent, how PIP2 can control so many different processes although it constitutes only about 1% of phospholipids of the plasma membrane. In general, there have been different hypotheses about the organization of PIP2 in cellular membranes. Since PIP2 is mainly found at the plasma membrane the dimensionality is reduced, thereby achieving an increase in concentration (McLaughlin and Murray, 2005). Additionally different pools of PIP2 are thought to exist. PIP2 is organized into clusters with local steep gradients within the plasma membrane. The clustering of two negatively charged PIP2 molecules is however energetically unfavorable (Wang et al., 2004) and the polyunsaturated hydrocarbon chain of PIP2 is in fact unlikely to spontaneously accumulate into rafts (McLaughlin et al., 2002). Since positively charged amino acids of basic proteins were shown to bind to several PIP2 molecules at the same time, these proteins could induce the cluster formation of PIP2 (Glaser et al., 1996; Laux et al., 2000). Through binding of basic proteins to PIP2 the lateral diffusion is reduced and thereby sequestration of other lipids takes place (Golebiewska et al., 2006). The reduced diffusion rates of PIP2 lead to an accumulation of more PIP2 around PIP2 clusters. This process thereby culminates in a different PIP2 turnover in these clusters from the bulk of the membrane PIP2 (Golub and Caroni, 2005). In agreement with this, non-caveolar cholesterol enriched rafts seem to have an enhanced mol fraction of PIP2 (Pike, 2004). The cluster formation of PIP2 molecules through basic proteins has thus been proposed as an initial signal to lead to the formation of rafts (McLaughlin and Murray, 2005; Golebiewska et al., 2006). Several membrane-associated proteins that bind PIP2 were shown to induce this accumulation of lipids into rafts (Golebiewska et al., 2006). This mechanism was proposed for PIPmodulins such as MARCKS, Gap43 and CAP23. MARCKS e.g. sequesters three PIP2 molecules per basic cluster (Fig. 4.1; Laux et al., 2000; McLaughlin

4 Discussion

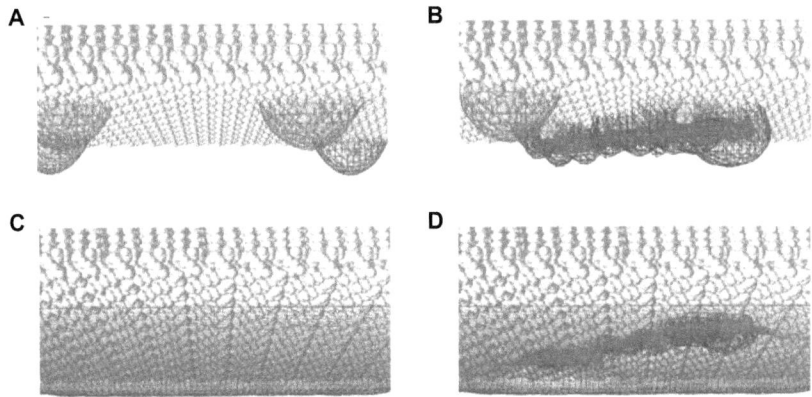

Figure 4.1: Lateral sequestration model modified from Golebiewska et al., 2006. (A) PIP2 molecules are distributed homogeneously within the plasma membrane consisting of phosphatidylcholine (PC) : PIP2 (99:1). (B) Upon binding of a positively charged protein, such as MARCKS or MBP, which electrostatically sequesters PIP2 molecules, PIP2 gets clustered into microdomains. (C) The surface potential adjacent to 2:1 PC : PS membrane at a distance of 1 nm from the surface is estimated as equipotential of -25 mV. (D) It was calculated that the binding of MARCKS to negative membranes induces a local positive potential of 25 mV, leading to an electrostatic trap that attracts PIP2 molecules and induces their clustering even if their physiological mol fraction is 100-fold lower than PS.

et al., 2002; Gambhir et al., 2004).

PIP2 gradients have been proposed to enable the signaling potential of PIP2, and are thought to form through both local synthesis by PIP-kinases and lateral sequestration by basic proteins (McLaughlin et al., 2002). A feedback loop in the activation of PIP2 generating enzymes together with a lateral sequestration of PIP2 then induces the steep gradients. Previous work suggested that the interaction of MBP with the plasma membrane is essential for the reorganization of membrane components that occurs when oligodendrocytes come in contact with neuronal processes. These studies have shown that MBP increases the lipid packaging of the myelin-forming membrane bilayer in cultured oligodendrocytes (Fitzner et al., 2006). The results presented here, provide evidence that the critical interaction of MBP is with the signaling lipid PIP2. The binding of basic proteins, i.e. MBP, to PIP2 might reduce the lateral diffusion of this lipid and also sequester other lipids into membrane microdomains. These domains may then act as recruitment sites for myelin membrane lipids, finally resulting in a condensation of the myelin membrane as determined by the fluorescent probe Laudran and two-photon microscopy (Fitzner et al., 2006) and the specification of myelin. In fact, results presented here indicate, that PIP2 and PIP3 might play crucial roles in defining the myelin membrane. PIP2 cluster might serve as a signalling platform, whereas PIP3, like e.g. during axon specification or during cell movement might induce accumulation of polarization factors at the membrane (Insall and Weiner, 2001). In fact, recent work in our lab suggested that loss of the PIP3 phosphatase PTEN induces hypermyelination in mice (Goebbels et al., in review). Additionally, recent findings confirm our hypothesis, that the clustering of lipids in myelin is indeed induced by the interaction of MBP with PIP2 (Musse et al., 2008). Since MBP is a basic protein with several positively charged amino acids distributed homogeneously along its sequence, it is likely that several PIP2 molecules interact with MBP (like MARCKS and other PIPmodulins). The interaction of MBP with PIP2 would therefore lead to the formation of PIP2 microdomains and possibly to the formation of lipid-ordered domains. In accordance with this hypothesis, clusters of MBP in membrane sheets were found, similar to previously observed PIP2 clusters (Milosevic et al., 2005; Fig. 3.5). Since there is no other basic protein present in myelin in such high concentrations to compensate for the loss of MBP in shiverer mice, shiverer oligodendrocytes loose the ability to cluster PIP2 and to form compaction of lipids (Fitzner et al., 2006).

4.2.3 Alteration of charges induces loss of compaction and binding of MBP

4.2.3.1 Alteration of membrane charge

The "classical" function of MBP is the formation of the major dense line (MDL) in compact myelin, since its absence in the CNS of shiverer mice show loss of compaction. This concept was confirmed by the discovery that the MDL in PNS myelin is dependent on myelin protein zero (P0), a single span adhesion protein with a highly basic intracellular domain (Privat et al., 1979; Lemke and Axel, 1985; Martini et al., 1995). The inner membrane surface potential, i.e. the potential created by negatively charged lipids and their counter ions attracted

4 Discussion

by them, has been estimated to be about 10^5V/cm (Olivotto et al., 1996). PIP2 contributes significantly to the surface charge having a valence of -4 at physiological pH (McLaughlin et al., 2002). Since MBP interacts with the membrane mostly through electrostatic interaction (Harauz et al., 2004), the hydrolysis of the most negatively charged lipid (PIP2) induced the dissociation of MBP from the plasma membrane. Both ATP depletion and ionomycin treatment lead to dissociation of MBP (Fig. 3.10). Our results therefore indicate that the reduction of the negative surface charge leads to dissociation of MBP from the plasma membrane, which indicates that MBP might partially regulate the surface potential. Interestingly, after dissociation from the plasma membrane, MBP was found in intracellular membranes that contain PS (but not PIP2). This result shows that MBP binds to different phospholipids, but preferentially to PIP2 possibly due its valence of -4 as compared to -1 of PS. I also note that the effect of ionomycin-induced MBP dissociation from the plasma membrane was more pronounced than the specific hydrolysis of PIP2 through synaptojanin 1 overexpression. This could be due to the additional flipping of PS from the inner leaflet of the plasma membrane to the outer leaflet, which occurs after Ca^{2+} influx (Bevers et al., 1983). Importantly, in acute brain slices that included white matter, ionomycin treatment and ATP depletion also led to a rapid myelin vesiculation (Fig. 3.12). Such a vesiculation would be expected if MBP serves to neutralize many of the negative surface charges of the closely interacting myelin membrane layers.

Therefore, I hypothesize here, that the mechanism by which the myelin sheath is compacted, maybe in part due reduced surface charge on the intracellular side of the membrane. Note that compact myelin is devoid of actin cytoskeleton, another player in structuring surface of cell membrane (Raucher et al., 2000). The expression of MBP coincides with a reduction of actin in the myelin membrane (unpublished observation). MBP is highly concentrated in compact myelin and its mRNA is directed to the plasma membrane, the site of its translation. The binding of MBP to the negatively charged membrane would reduce the negativity of the membrane and would therefore induce compaction of the two opposing myelin membranes. In a spherical cell, the negatively charged lipids on the inside of the cell surface produce an internal field and due to mutual repulsion create a tension on the surface of the cell. Counterions or proteins containing a positive charge, present in the cell cytoplasm are likely to reduce this effect. Proteins that bind to negatively charged lipids

4 Discussion

such as PIP2 would therefore lead to a reduction of the surface charge. The membrane would therefore collapse onto the other membrane, leading to compaction (Smith, 1977; Inouye and Kirschner, 1988b).

It has previously been suggested that the compaction of cytosolic surfaces through MBP might be sensitive to changes in surface charge (Inouye and Kirschner, 1988a; Boggs, 2006). These fluctuations might involve neuronally regulated Ca^{2+} influx or pH changes in oligodendrocytes (Ro and Carson, 2004) and might therefore regulate MBP mediated compaction, as suggested in other cell types (McLaughlin, 1989). However, calcium/CAM might also influence the association of MBP to the plasma membrane, since intracellular Ca^{2+} levels were shown to regulate the binding of MBP to Calmodulin (Boggs, 2006). Additionally, other basic unstructured proteins such as MARCKS, were shown to be released from the plasma membrane in a Ca/CAM dependent manner.

Abnormal Ca^{2+}-entry is a frequent sign of cellular pathology. Thus, the effects of elevated intracellular Ca^{2+} on MBP-membrane interactions may be relevant to the changes of myelin that occur under various pathological conditions. Recent studies have shown that CNS myelin contains NMDA receptors that could be responsible for a rise in intracellular Ca^{2+} when the white matter is injured by hypoxia or excitotoxicity (Káradóttir et al., 2005; Micu et al., 2006). Moreover, intracellular accumulation of Ca^{2+} induced by glycine/glutamate signaling, was shown to disrupt the myelin ultrastructure. In addition, it has been reported that oxygen-glucose deprivation causes a similar vesiculation following the activation of AMPA/kainate receptors (Micu et al., 2006; Tekkök et al., 2005).

The previous notion that myelin is an inert membrane, whose lipid and protein turnover is diminishable has changed (reviewed in Ledeen, 1984). Several lipid metabolizing enzymes were found active in myelin. Phosphoinositides were shown to undergo relatively high turnover and highly concentrated in myelin (1.5%) compared to normal plasma membrane (1%) (Deshmukh et al., 1981; Kahn and Morell, 1988). It has been reported that PIP and PIP2 turnover in myelin might involve PO^{3-} derived from axons (Chakraborty et al., 1999). Since the turnover rate of PIP2 is relatively high, ATP depletion leads to a loss of PIP2 in myelin, MBP dissociation and myelin vesiculation. Previous studies had indicated that PI delivery might be dependent on neuronal transport of PO^{3-}. The ex vivo-experiments showed that vesiculation was mainly observed at the inner loop, possibly due to two reasons:

4 Discussion

first, the cytoplasm is localized in the inner loop and MBP can therefore dissociate from the plasma membrane. Second, if PI is delivered from the axon to the myelin, the inner loop being closest to the axon might be most sensitive to PI changes. The rest of myelin might still have sufficient amount of PIP2 trapped, so that the compaction is still preserved. However, incubation for longer times might lead to the loss of myelin compaction in all layers of myelin. Since only part of myelin is non-compacted, I was not able to show that vesiculation is due to MBP dissociation from the membrane through biochemical methods. However, pre-incubation of neomycin prevented vesiculation. Although neomycin was also shown to alter the surface charge (Gabev et al., 1989), neomycin-blockage was sufficient to prevent vesiculation after 30 min of incubation with ionomycin, but not after 1h (Fig. 3.13 and not shown). It is possible that PIP2 lipids can bind both MBP and neomycin at the same time. Thereby, neomycin-bound PIP2 molecules are shielded from PLC-mediated hydrolysis.

X-ray diffraction measurements have shown that the spacing between myelin lamellae increased with increasing ionic strength and pH (Inouye and Kirschner, 1988a). Additionally these results indicated that myelin from shiverer mice exhibited stronger sensitivity to ionic strength than myelin from wild-type mice. The surface charge density which influences the periodicity was increased in shiverer PNS myelin, again indicating an influence of MBP in blocking the negative surface charge. The periodicity was assumed to be dependent on an interplay between repulsive forces such as the electrostatic repulsion force (surface potential), hydration force and undulation force (proposed through undulation of lipid membranes) and van-der Waals attractive force. Therefore the distance between the two membranes is influenced by a variety of forces that are dependent on pH and ionic strength as well as the available negative phospholipids present on the surface. In their study it was noted that the surface charge density however is dependent on the distribution of proteins, since PNS and CNS myelin periodicity responded differently towards changes in pH or ionic strength (Inouye and Kirschner, 1988a). Proteins known to play a role in compaction of myelin such as PLP and MBP in the CNS and P0 and MBP in the PNS are therefore supposed to influence the energy needed to bring the two opposing membranes in such close apposition.

4.2.3.2 Reduction of charges in MBP and its effect on membrane association

Since MBP interacts with the plasma membrane mostly through electrostatic interactions, alteration of charges of the membrane or the protein should lead to an alteration in binding of the protein. Experiments with different deletion mutants of recombinant MBP have indicated the importance of N and C terminal domain in the organization of lipid surfaces (Hill et al., 2003). I generated a truncated form of MBP14k by removing exon 1 or exon 7-encoded region, in order to investigate which region of MBP is sufficient to associate to the plasma membrane. In fact, the truncated N-terminal part of MBP (encoded in exon-1) was sufficient to bind to the plasma membrane, whereas the C-terminal part of MBP (encoded in exon 7) was not. However, one has to take into account that both constructs were fused to YFP at the C-terminal part of MBP. Therefore, it is possible that the exon-7-encoded YFP fusion protein might not bind to the plasma membrane, because YFP might prevent the association to it. Additional replacement of different positively charged amino acids with Ala in exon 1-encoded region, led to a loss of binding to the plasma membrane. However, when R10 and K12 were both exchanged for Ala, it had no effect on plasma membrane localization. These experiments show that exon1-encoded region of MBP binds with less affinity to the plasma membrane, compared to full length MBP14k. Hence, reducing the number of positive amino acids reduced the binding capacity of MBP. These results also indicate that the N-terminal part of MBP (encoded in exon 1) is sufficient to bind to the plasma membrane. Additionally, not only the number of positive charges controls the association of MBP to the plasma membrane, but also their position. The tertiary structure of MBP might therefore also influence the efficiency to associate with the plasma membrane.

Most posttranslational modifications reduce the overall positive charge of MBP. Several studies have indicated that these posttranslational modifications of MBP influence its binding affinity to the membrane (Boggs et al., 1997). Since an increased amount of citrullinated MBP is found in Multiple Sclerosis patients, this indicates that the overall charge of MBP seems to be important for its proper function (Kim et al., 2003). Additionally, less positively charged isomers have a reduced ability to organize lipid membranes (Shanshiashvili et al., 2003). In fact, studies have shown that the binding energy of basic proteins to PIP2 increases linearly with the number of basic residues and therefore depends on the overall

4 Discussion

charge of the protein (Kim et al., 1991; Ben-Tal et al., 1996). In their study Kim et al. used peptides composed of various amounts of Lys or Arg and assessed their binding affinity to acidic membranes. In their model, each addition of a basic amino acid decreased the concentration of peptide that is needed to reverse the charge of phosphatidylserine or phosphatidylglycerol vesicles by a factor of ten and each basic residue added to the peptide increased the binding affinity ten fold (Kim et al., 2003). They conclude that in proteins with polybasic clusters each binding of basic residues to acidic lipids induces the next basic residue to bind to it. Since the basic amino acids within MBP are distributed along the sequence rather than clustered into domains, it is possible that it can shield more effectively the negatively charged lipids, than if these amino acids were clustered.

Previous in vitro data had suggested a covalent linkage of MBP with PIP2 (Chang et al., 1986; Yang et al., 1986). These studies had revealed that the sequence G-S-G-K probably binds to PIP2 and that S54 is covalently attached to it. Our site directed mutagenesis studies show however that mutant MBP (S54A) was less bound to the plasma membrane compared to wild-type MBP, but not sufficient to loose plasma membrane association. In conclusion, these experiments indicate that MBP probably interacts with the membrane mainly through electrostatic interactions, but that MBP might also be linked to PIP2 through a covalent or hydrophobic interaction with S54.

Protein adhesion and force-distance measurements showed that both the charges of the membrane as well as of protein, that binds with opposite charges to it, determines the repulsive forces of the two opposing membranes. Small changes of any of the two parameters can lead to changes in adhesion of myelin membrane (Hu et al., 2004). Our results confirm that both parameters are necessary for the binding of MBP: Our mutagenesis study had indicated that alteration of overall charges of MBP reduces its membrane association as well as our experiments on surface charge alteration.

4.2.4 Possible Involvement of MBP in process outgrowth and alteration in membrane tension

It is known that purinergic signalling or Glutamate receptor activation leads to a rise in intracellular Ca^{2+} in oligodendrocytes (Holzwarth et al., 1994; Liu et al., 1997; Stevens et al., 2002). However, it would be surprising if extracellular neuronal signals lead to a rise in Ca^{2+} in oligodendrocytes, which would result in the dissociation of MBP from the plasma membrane, since during differentiation of oligodendrocytes MBP accumulates at membrane sheets. A possible model for this discrepancy was proposed by Sheetz for the basic protein MARCKS (Sheetz et al., 2006). During polarization the actin cytoskeleton is instable (Bradke and Dotti, 1999). MARCKS associates with the plasma membrane and leads to the accumulation of PIP2. External stimuli then lead to the dissociation of MARCKS due to influx of Ca^{2+} and to the activation of PLC, which in turn leads to the formation of F-actin at the growing processes. Recruitment of type I PIPK then leads to the accumulation of PIP2, which then directs MARCKS back to the plasma membrane. Additionally, it was shown that PIP2 levels regulate the adhesion between membrane and cytoskeleton and therefore the membrane tension (Raucher et al., 2000). Considering the structure of myelin or the initial cup formed around an axon, one can imagine that for such a structure to form, the membrane has to bend. For this purpose, either membrane tension is increased, leading to bending, or membrane tension is decreased due to excess membrane flow enabling bending of the membrane. In fact, it was shown that during phagocytosis the membrane tension is reduced (Herant et al., 2006). The initial step of myelination might be similar to initial cup formation during phagocytosis. Since it was shown that PIP2 regulates membrane tension through its interaction with actin binding proteins, it might be possible that the interaction of PIP2 with MBP plays a role in regulating membrane tension. Additionally myelin is devoid of F-actin, which regulates cell shape and membrane tension. MBP might therefore replace the role of actin in regulating membrane tension within compact myelin. MBP dissociated from the plasma membrane in response to Ca^{2+} influx, similar to GFP-PH-PLCδ1. GFP-PH-PLCδ1 has previously been used to reduce membrane tension, since it can block the interaction of PIP2 with actin-binding proteins. Second, when the surface charge was altered, the dissociation of PH-PLCδ1 or MBP was followed by a sudden increase in processes. Process outgrowth was also shown to involve alteration of membrane tension. Additionally, it was suggested that ionomycin treatment induces actin depolymerization (Sogabe et al., 1996). Since depolarization of cells did not induce dissociation of MBP, Ca^{2+} influx seems to specifically induce this effect. However, activation of some unknown receptors might induce Ca^{2+} influx at specific sites and process outgrowth. Similar to PH-PLCδ1, the binding of MBP to PIP2 might lead to a decrease in membrane tension, since MBP could block the binding of actin binding proteins to PIP2.

The influence of MBP in membrane dynamics seems complex. These experiments were not sufficient to illustrate whether MBP plays any regulatory role, since the results are still not conclusive. More elaborate experiments such as atomic-force microscopy or laser optical tweezer experiments have to be used to answer such questions. It is possible that the dynamics of membrane outgrowth is driven by the association and subsequent dissociation of MBP to and from the membrane. Binding of MBP leads to a decrease of membrane tension, Ca^{2+} influx leads to MBP dissociation dependent on PLC activation and to a subsequent membrane outgrowth, as proposed by Sheetz (Sheetz et al., 2006). MBP might therefore be involved in the fine-tuning and regulation of membrane tension, by altering the available pool of PIP2 at the plasma membrane.

Domains that were previously shown to interact with PIP2 and influence membrane curvature are the so-called BAR domains (Zimmerberg and McLaughlin, 2004). Proteins that contain a BAR domain were shown to interact with PIP2 at the plasma membrane and were shown to induce membrane curvature due to their bent secondary structure (Zimmerberg and McLaughlin, 2004). MBP as a basic protein might also act like a BAR domain-containing protein, influencing membrane curvature (Boggs et al., 2006). It is intriguing to note that since shiverer mice have no myelin, MBP is likely to play a role in its formation not only in its compaction. The finding that MBP interacts with PIP2 opens many interesting questions. If MBP is involved in regulating the available pool of PIP2, it might be involved in exo- and endocytosis or in general membrane outgrowth.

4.3 Conclusion

This study has provided experimental evidence that PIP2, a highly charged phospholipid, is an essential component for the binding of MBP to the cell membrane, and in extension for the function of MBP in myelination. In a loss of function experiment, MBP dissociated from the plasma membrane upon Ca^{2+} influx or specific PIP2 hydrolysis (through synaptojanin1 expression). On the other hand, PIP2 accumulation in endomembranes induced a relocalization of MBP to these endomembranes. Reduction of the overall charge of MBP reduced its binding to the plasma membrane. One critical binding domain, harboring critical positively charged residues, was found in the exon1-encoded region of MBP. Moreover, reduction of surface charge induced loss of MBP-plasma membrane association and induced myelin vesiculation in acute brain slices. It is intriguing to note that PIP2 has independently from this study been implicated in myelination as a signaling lipid and substrate of PI3K, which generates PIP3 and activates the AKT-pathway to drive CNS myelination (Flores et al., 2008; Goebbels et al., in preparation). PIP3 might play a role in polarization of oligodendrocytes, presumably as a response to specific axonal signals. It is possible that oligodendroglial PIP2 is equally important as a signaling lipid and as a lipid docking site for MBP to fulfill its function in lipid sorting and myelin compaction. Interfering with this highly specific lipid-protein interaction, for example by abnormal increases of intracellular Ca^{2+}, leads to the destabilization of CNS myelin and may be related to myelin destruction in ischemic conditions and in demyelinating diseases.

5 Supplemental material

Biochemical quantification of plasma membrane localization of MBP after PIP2 depletion

Additional to FRET, membrane sheets and other microscopic techniques, I wanted to confirm the PIP2 dependent interaction of MBP with the plasma membrane through biochemical methods. To specifically reduce PIP2 levels in cells, Oli-neu cells were transfected with MBP14k-YFP and mRFP-Synj1 contructs or control vector respectively. In order to isolate MBP from the plasma membrane we used various previously described methods (Simons et al., 2000). As mentioned before, MBP is localized in detergent resistant membranes (DRM) within the plasma membrane of oligodendrocytes. Therefore, after cell lysis samples were incubated with detergent (20 mM CHAPS). Membrane was separated from the cytosol through flotation on Optiprep gradient (Fitzner et al., 2006). Fig. 5.1A shows the different fractions. The isolation of DRM however showed no obvious difference in localization of MBP between samples taken from Synj1 expressing cells or control cells. Instead of isolating only the DRM fraction of the plasma membrane, I therefore isolated the complete plasma membrane by centrifuging the cell lysate at 100000g (Fig. 5.1B). The pellet fraction represents the plasma membrane. This method however did not show any obvious difference. Also different modification of the lysis buffer did not result in a difference between the two probes. Additionally, flotation of the membrane fraction on a sucrose gradient (Fig. 5.2C) also did not show any obvious quantitative difference in membrane localization of MBP.

One of the possible explanations for this is that the amount of co-transfected cells was not enough to quantitatively measure the reduction of MBP at the plasma membrane. Additionally, the amount of protein isolated with these procedures was not enough to quan-

5 Supplemental material

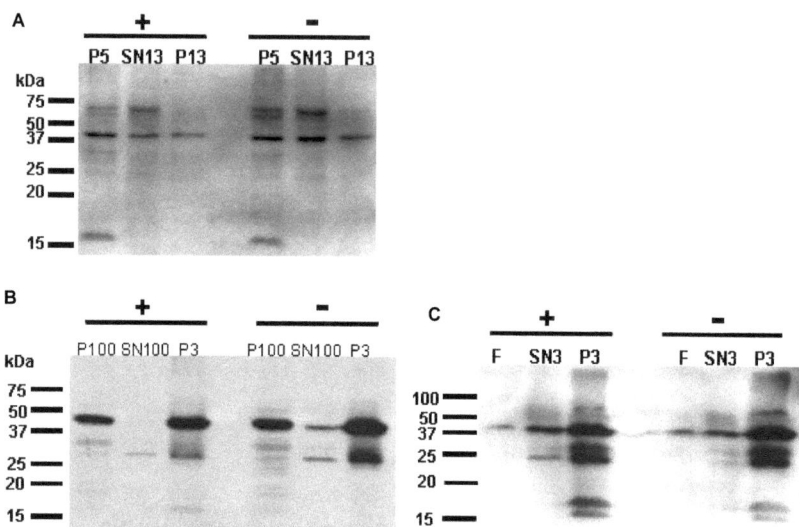

Figure 5.1: Different biochemical methods were used to separate plasma membrane from cytosol of Oli-neu cells transfected either with MBP14k-YFP and mRFP-Synj1 or control vector respectively.
(A) DRMs were isolated as described in Methods. First post-nuclear fraction was separated (P5) before centrifuging the superntant at 13000 rpm. The detergent resistant membrane fraction P13 compared to the rest does not show any obvious difference.
(B) Complete plasma membrane isolation through centrifugation at 100000g also did not show any difference. (P3= postnuclear fraction; P100= membrane fraction; SN100= cytosolic fraction); (C) Flotation of membrane on sucrose gradient also did not result in a difference between the two samples. (F= flotated membrane; SN3= postnuclear fraction; SN3= postnuclear supernatant)

titatively measure a difference between MBP concentrated at the plasma membrane compared to the cytosol. I therefore decided to isolate recombinant MBP from transduced E.Coli strands. One possible way of measuring plasma membrane association is through Isothermal-Calomery using different phospholipids as binding partners.

Generation of stable cell-lines expressing MBP14k-YFP and MBP21k-YFP

As a tool for our investigations we decided to stably transfect Oli-neu cells as well as OLN93 cells with MBP14k-YFP and MBP21k-YFP constructs. Cells were grown to about 60% confluency and transfected with respective linearized plasmids. Clones were selected through addition of hygromycin. Surprisingly both isoforms were found in the nucleus rather than at the plasma membrane in both cell lines, although only 21kDa MBP isoform had previously been reported to be localized to the nucleus (Fig. 5.2; Pedraza et al., 1997).

5 Supplemental material

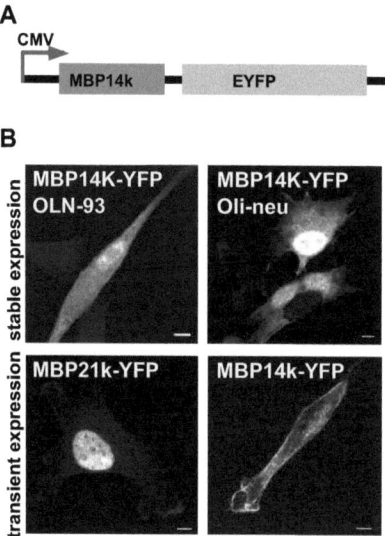

Figure 5.2: Generation of stable cell-lines expressing MBP14k-YFP and MBP21k-YFP. As a tool we stably transfected Oli-neu and OLN-93 oligodendrocyte precursor cell lines with MBP14k-YFP or MBP21-YFP constructs. (A) Both genes were driven by a CMV promotor and were tagged with YFP at the C-terminus of MBP. (B) In both cell types however, MBP was mostly localized to the nucleus and was not targeted to the plasma membrane (scale bar 5 μm).

References

Ainger K, Avossa D, Diana AS, Barry C, Barbarese E, Carson JH (1997) Transport and Localization Elements in Myelin Basic Protein mRNA. J. Cell Biol. 138:1077-1087

Anderson RA, Boronenkov IV, Doughman SD, Kunz J, Loijens JC (1999) Phosphatidylinositol phosphate kinases, a multifaceted family of signaling enzymes. J. Biol. Chem. 274:9907-9910

Arbuzova A, Martushova K, Hangyás-Mihályné G, Morris AJ, Ozaki S, Prestwich GD, McLaughlin S (2000) Fluorescently labeled neomycin as a probe of phosphatidylinositol-4, 5-bisphosphate in membranes. Biochim. Biophys. Acta 1464:35-48

Banker G (2003) Pars, PI 3-kinase, and the Establishment of Neuronal Polarity. Cell 112:4-5

Bansal R, Warrington AE, Gard AL, Ranscht B, Pfeiffer SE (1989) Multiple and novel specificities of monoclonal antibodies O1, O4, and R-mAb used in the analysis of oligodendrocyte development. J. Neurosci. Res. 24:548-557

Barbarese E, Brumwell C, Kwon S, Cui H, Carson JH (1999) RNA on the road to myelin. J. Neurocytol. 28:263-270

Baron W, Decker L, Colognato H, ffrench-Constant C (2003) Regulation of integrin growth factor interactions in oligodendrocytes by lipid raft microdomains. Curr. Biol. 13:151-155

Bates IR, Boggs JM, Feix JB, Harauz G (2003) Membrane-anchoring and Charge Effects in the Interaction of Myelin Basic Protein with Lipid Bilayers Studied by Site-directed Spin Labeling. J. Biol. Chem. 278:29041-29047

Baumann N, Pham-Dinh D (2001) Biology of oligodendrocyte and myelin in the mammalian central nervous system. Physiol. Rev. 81:871-927

Ben-Tal N, Honig B, Peitzsch RM, Denisov G, McLaughlin S (1996) Binding of small basic peptides to membranes containing acidic lipids: theoretical models and experimental results. Biophys. J. 71:561-575

Bevers EM, Comfurius P, Zwaal RF (1983) Changes in membrane phospholipid distribution during platelet activation. Biochim. Biophys. Acta 736:57-66

Boggs JM, Rangaraj G (2000) Interaction of lipid-bound myelin basic protein with actin filaments and calmodulin. Biochemistry 39:7799-806

Boggs JM, Rangaraj G, Koshy KM, Ackerley C, Wood DD, Moscarello MA (1999) Highly deiminated isoform of myelin basic protein from multiple sclerosis brain causes fragmentation of lipid vesicles. J. Neurosci. Res. 57:5295-35

Boggs JM, Yip PM, Rangaraj G, Jo E (1997) Effect of posttranslational modifications to myelin basic protein on its ability to aggregate acidic lipid vesicles. Biochemistry 36:5065-5071

Boggs J (2006) Myelin basic protein: a multifunctional protein. Cell. Mol. Life Sci. (CMLS) 63:1945-1961

Bradke F, Dotti CG (1999) The role of local actin instability in axon formation. Science 283:1931-1934

Bremer C (2008) Optical Methods. In: Handb. Exp. Pharmacol. 185Pt: 3-12. Berlin Heidelberg: Springer

Brinkmann BG, Agarwal A, Sereda MW, Garratt AN, Müller T, Wende H, Stassart RM, Nawaz S, Humml C, Velanac V, Radyushkin K, Goebbels S, Fischer TM, Franklin RJ, Lai C, Ehrenreich H, Birchmeier C, Schwab MH, Nave KA (2008) Neuregulin-1/ErbB signaling serves distinct functions in myelination of the peripheral and central nervous system. Neuron 59:581-595

Brown FD, Rozelle AL, Yin HL, Balla T, Donaldson JG (2001) Phosphatidylinositol 4,5-bisphosphate and Arf6-regulated membrane traffic. J. Cell Biol. 154:1007-1017

Campagnoni AT and Campagnoni CW (2004) Myelin basic protein gene. In: Myelin biology and disorders (Lazzarini RA) pp 378-395. San Diego: Elsevier Academic Press

Carson JH, Nielson ML, Barbarese E (1983) Developmental regulation of myelin basic protein expression in mouse brain. Dev. Biol. 96:485-492

Chakraborty G, Drivas A, Ledeen R (1999) The phosphoinositide signaling cycle in myelin requires cooperative interaction with the axon. Neurochem. Res. 24:249-254

Chan JR, Jolicoeur C, Yamauchi J, Elliott J, Fawcett JP, Ng BK, Cayouette M (2006) The polarity protein Par-3 directly interacts with p75NTR to regulate myelination. Science 314:832-836

Chang PC, Yang JC, Fujitaki JM, Chiu KC, Smith RA (1986) Covalent linkage of phospholipid to myelin basic protein: identification of serine-54 as the site of attachment. Biochemistry 25:2682-2686

Cremona O, Di Paolo G, Wenk MR, Lüthi A, Kim WT, Takei K, Daniell L, Nemoto Y, Shears SB, Flavell RA, McCormick DA, De Camilli P (1999) Essential role of phosphoinositide metabolism in synaptic vesicle recycling. Cell 99:179-188

Czech MP (2000) PIP2 and PIP3: Complex Roles at the Cell Surface. Cell 100:603-606

Debruin LS, Harauz G (2007) White matter rafting–membrane microdomains in myelin. Neurochem. Res. 32:213-228

Demel RA, London Y, Geurts van Kessel WS, Vossenberg FG, van Deenen LL (1973) The specific interaction of myelin basic protein with lipids at the air-water interface. Biochim. Biophys. Acta 311:507-519

References

Deshmukh DS, Kuizon S, Bear WD, Brockerhoff H (1981) Rapid incorporation in vivo of intracerebrally injected 32Pi into polyphosphoinositides of three subfractions of rat brain myelin. J. Neurochem. 36:594-601

Di Paolo G, De Camilli P (2006) Phosphoinositides in cell regulation and membrane dynamics. Nature 443:651-657

Donaldson JG (2003) Multiple Roles for Arf6: Sorting, Structuring, and Signaling at the Plasma Membrane. J. Biol. Chem. 278:41573-41576

Doughman RL, Firestone AJ, Anderson RA (2003) Phosphatidylinositol Phosphate Kinases Put PI4,5P 2 in Its Place. J. Membr. Biol. 194:77-89

Eisenbarth GS, Walsh FS, Nirenberg M (1979) Monoclonal antibody to a plasma membrane antigen of neurons. Proc. Natl. Acad. Sci. U. S. A. 76:4913-4917

de Ferra F, Engh H, Hudson L, Kamholz J, Puckett C, Molineaux S, Lazzarini RA (1985) Alternative splicing accounts for the four forms of myelin basic protein. Cell 43:721-727

Fitzner D, Schneider A, Kippert A, Möbius W, Willig KI, Hell SW, Bunt G, Gaus K, Simons M (2006) Myelin basic protein-dependent plasma membrane reorganization in the formation of myelin. EMBO J. 25:5037-5048

Flores AI, Narayanan SP, Morse EN, Shick HE, Yin X, Kidd G, Avila RL, Kirschner DA, Macklin WB (2008) Constitutively active Akt induces enhanced myelination in the CNS. J. Neurosci. 28:7174-7183

Gabev E, Kasianowicz J, Abbott T, McLaughlin S (1989) Binding of neomycin to phosphatidylinositol 4,5-bisphosphate (PIP2). Biochim. Biophys. Acta 979:105-112

Gambhir A, Hangyás-Mihályné G, Zaitseva I, Cafiso DS, Wang J, Murray D, Pentyala SN, Smith SO, McLaughlin S (2004) Electrostatic sequestration of PIP2 on phospholipid membranes by basic/aromatic regions of proteins. Biophys. J. 86:2188-21207

Gamper N, Shapiro MS (2007a) Target-specific PIP(2) signalling: how might it work? J. Physiol. 582:967-975

Gamper N, Shapiro MS (2007b) Regulation of ion transport proteins by membrane phosphoinositides. Nat. Rev. Neurosci. 8:921-934

Glaser M, Wanaski S, Buser CA, Boguslavsky V, Rashidzada W, Morris A, Rebecchi M, Scarlata SF, Runnels LW, Prestwich GD, Chen J, Aderem A, Ahn J, McLaughlin S (1996) Myristoylated Alanine-rich C Kinase Substrate (MARCKS) Produces Reversible Inhibition of Phospholipase C by Sequestering Phosphatidylinositol 4,5-Bisphosphate in Lateral Domains. J. Biol. Chem. 271:26187-26193

Goldstein B, Macara IG (2007) The PAR Proteins: Fundamental Players in Animal Cell Polarization. Dev. Cell 13:609-622

Golebiewska U, Gambhir A, Hangyás-Mihályné G, Zaitseva I, Rädler J, McLaughlin S (2006) Membrane-bound basic peptides sequester multivalent (PIP2), but not monovalent (PS), acidic lipids. Biophys. J. 91:588-599

Golub T, Caroni P (2005) PI(4,5)P2-dependent microdomain assemblies capture microtubules to promote and control leading edge motility. J. Cell Biol. 169:151-165

Grand RJ, Perry SV (1980) The binding of calmodulin to myelin basic protein and histone H2B. Biochem. J. 189:227-240

Haas H, Steitz R, Fasano A, Liuzzi GM, Polverini E, Cavatorta P, Riccio P (2007) Laminar order within Langmuir-Blodgett multilayers from phospholipid and myelin basic protein: a neutron reflectivity study. Langmuir 23:8491-8496

Hall A, Giese NA, Richardson WD (1996) Spinal cord oligodendrocytes develop from ventrally derived progenitor cells that express PDGF alpha-receptors. Development 122:4085-4094

Harauz G, Ishiyama N, Hill CMD, Bates IR, Libich DS, Farés C (2004) Myelin basic protein-diverse conformational states of an intrinsically unstructured protein and its roles in myelin assembly and multiple sclerosis. Micron 35:503-542

Heo WD, Inoue T, Park WS, Kim ML, Park BO, Wandless TJ, Meyer T (2006) PI(3,4,5)P3 and PI(4,5)P2 lipids target proteins with polybasic clusters to the plasma membrane. Science 314:1458-1461

Herant M, Heinrich V, Dembo M (2006) Mechanics of neutrophil phagocytosis: experiments and quantitative models. J. Cell Sci. 119:1903-1913

Hilgemann DW, Feng S, Nasuhoglu C (2001) The complex and intriguing lives of PIP2 with ion channels and transporters. Sci STKE 2001:RE19

Hill CMD, Haines JD, Antler CE, Bates IR, Libich DS, Harauz G (2003) Terminal deletion mutants of myelin basic protein: new insights into self-association and phospholipid interactions. Micron 34:25-37

Holzwarth JA, Gibbons SJ, Brorson JR, Philipson LH, Miller RJ (1994) Glutamate receptor agonists stimulate diverse calcium responses in different types of cultured rat cortical glial cells. J Neurosci 14:1879-91

Hu Y, Doudevski I, Wood D, Moscarello M, Husted C, Genain C, Zasadzinski JA, Israelachvili J (2004) Synergistic interactions of lipids and myelin basic protein. Proc. Natl. Acad. Sci. U.S.A. 101: 13466Ð13471.

Hu Y, Israelachvili J (2008) Lateral reorganization of myelin lipid domains by myelin basic protein studied at the air-water interface. Colloids Surf B Biointerfaces 62:22-30

Inouye H, Kirschner DA (1988a) Membrane interactions in nerve myelin: II. Determination of surface charge from biochemical data. Biophys. J. 53:247-260

Inouye H, Kirschner DA (1988b) Membrane interactions in nerve myelin: I. Determination of surface charge from effects of pH and ionic strength on period. Biophys. J. 53:235-245

Insall RH, Weiner OD (2001) PIP3, PIP2, and Cell Movement–Similar Messages, Different Meanings? Dev. Cell 1:743-747

Jessen KR (2004) Glial cells. Int. J. Biochem. Cell Biol. 36:1861-1867

Jung M, Krämer E, Grzenkowski M, Tang K, Blakemore W, Aguzzi A, Khazaie K, Chlichlia

References

K, von Blankenfeld G, Kettenmann H (1995) Lines of murine oligodendroglial precursor cells immortalized by an activated neu tyrosine kinase show distinct degrees of interaction with axons in vitro and in vivo. Eur. J. Neurosci. 7:1245-1265

Kahn DW, Morell P (1988) Phosphatidic acid and phosphoinositide turnover in myelin and its stimulation by acetylcholine. J. Neurochem. 50:1542-1550

Káradóttir R, Cavelier P, Bergersen LH, Attwell D (2005) NMDA receptors are expressed in oligodendrocytes and activated in ischaemia. Nature 438:1162-116

Karlsson U, Schultz RL (1965) Fixation of the central nervous system from electron microscopy by aldehyde perfusion. I. Preservation with aldehyde perfusates versus durect perfusion with osmium tetroxide with special preference to membranes and the extracellular space. J. Ultrastruct. Res 12:160-86

Kim J, Mosior M, Chung LA, Wu H, McLaughlin S (1991) Binding of peptides with basic residues to membranes containing acidic phospholipids. Biophys. J. 60:135-148

Kim JK, Mastronardi FG, Wood DD, Lubman DM, Zand R, Moscarello MA (2003) Multiple sclerosis: an important role for post-translational modifications of myelin basic protein in pathogenesis. Mol. Cell Proteomics 2:453-462

Kimura M, Sato M, Akatsuka A, Nozawa-Kimura S, Takahashi R, Yokoyama M, Nomura T, Katsuki M (1989) Restoration of myelin formation by a single type of myelin basic protein in transgenic shiverer mice. Proc. Natl. Acad. Sci. U.S.A. 86:5661-5665

Kippert A, Trajkovic K, Rajendran L, Ries J, Simons M (2007) Rho Regulates Membrane Transport in the Endocytic Pathway to Control Plasma Membrane Specialization in Oligodendroglial Cells. J. Neurosci. 27:3560-3570

Kirby BB, Takada N, Latimer AJ, Shin J, Carney TJ, Kelsh RN, Appel B (2006) In vivo time-lapse imaging shows dynamic oligodendrocyte progenitor behavior during zebrafish development. Nat Neurosci 9:1506-1511

Kleitman N, Wood PM, Bunge RP (1998) Tissue culture methods for the study of myelination. In: Culturing Nerve Cells pp545-640. The MIT Press

Krauß M, Haucke V (2007) Phosphoinositide-metabolizing enzymes at the interface between membrane traffic and cell signalling. EMBO Rep. 8:241Ð246

Lang T, Bruns D, Wenzel D, Riedel D, Holroyd P, Thiele C, Jahn R (2001) SNAREs are concentrated in cholesterol-dependent clusters that define docking and fusion sites for exocytosis. EMBO J. 20:2202-2213

Laux T, Fukami K, Thelen M, Golub T, Frey D, Caroni P (2000) GAP43, MARCKS, and CAP23 modulate PI(4,5)P(2) at plasmalemmal rafts, and regulate cell cortex actin dynamics through a common mechanism. J. Cell Biol. 149:1455-1472

Ledeen RW (1984) Lipid-metabolizing enzymes of myelin and their relation to the axon. J. Lipid Res. 25:1548-1554

Lee AG (2001) Myelin: Delivery by raft. Curr. Biol. 11:R60-62

Lemke G, Axel R (1985) Isolation and sequence of a cDNA encoding the major structural protein of peripheral myelin. Cell 40:501-508

Lemmon MA, Ferguson KM, O'Brien R, Sigler PB, Schlessinger J (1995) Specific and high-affinity binding of inositol phosphates to an isolated pleckstrin homology domain. Proc. Natl. Acad. Sci. U. S. A. 92:10472-10476

Liu C, Hermann TE (1978) Characterization of ionomycin as a calcium ionophore. J. Biol. Chem. 253:5892-5894

Liu HN, Molina-Holgado E, Almazan G (1997) Glutamate-stimulated production of inositol phosphates is mediated by Ca2+ influx in oligodendrocyte progenitors. Eur. J. Pharmacol. 338:277-287

Lowden JA, Moscarello MA, Morecki R (1966) The isolation and characterization of an acid-soluble protein from myelin. Can. J. Biochem. 44:567-577

Lubetzki C, Demerens C, Anglade P, Villarroya H, Frankfurter A, Lee VM, Zalc B (1993) Even in culture, oligodendrocytes myelinate solely axons. Proc. Natl. Acad. Sci. U. S. A. 90:6820-6824

Maier O, Hoekstra D, Baron W (2008) Polarity development in oligodendrocytes: sorting and trafficking of myelin components. J. Mol. Neurosci. 35:35-53

Martin-Belmonte F, Gassama .A, Datta A, Yu W, Rescher U, Gerke V, Mostov K (2007) PTEN-Mediated Apical Segregation of Phosphoinositides Controls Epithelial Morphogenesis through Cdc42. Cell Vol 128:383-397

Martini R, Zielasek J, Toyka KV, Giese KP, Schachner M (1995) Protein zero (P0)-deficient mice show myelin degeneration in peripheral nerves characteristic of inherited human neuropathies. Nat. Genet. 11:281-286

McLaughlin S (1989) The electrostatic properties of membranes. Annu. Rev. Biophys. Biophys. Chem. 18:113-136

McLaughlin S, Murray D (2005) Plasma membrane phosphoinositide organization by protein electrostatics. Nature 438:605-611

McLaughlin S, Wang J, Gambhir A, Murray D (2002) PIP(2) and proteins: interactions, organization, and information flow. Annu. Rev. Biophys. Biomol. Struct. 31:151-75

Mellman I (2000) Quo vadis: polarized membrane recycling in motility and phagocytosis. J. Cell Biol. 149:529-530

Ménager C, Arimura N, Fukata Y, Kaibuchi K (2004) PIP3 is involved in neuronal polarization and axon formation. J. Neurochem. 89:109-118

Micu I, Jiang Q, Coderre E, Ridsdale A, Zhang L, Woulfe J, Yin X, Trapp BD, McRory JE, Rehak R, Zamponi GW, Wang W, Stys PK (2006) NMDA receptors mediate calcium accumulation in myelin during chemical ischaemia. Nature 439:988-992

Milosevic I, S¿rensen JB, Lang T, Krauss M, Nagy G, Haucke V, Jahn R, Neher E (2005) Plasmalemmal phosphatidylinositol-4,5-bisphosphate level regulates the releasable vesicle pool size in chromaffin cells. J. Neurosci. 25:2557-2565

References

Modesti NM, Barra HS (1986) The interaction of myelin basic protein with tubulin and the inhibition of tubulin carboxypeptidase activity. Biochem. Biophys. Res. Commun. 136:482-489

Molineaux SM, Engh H, de Ferra F, Hudson L, Lazzarini RA (1986) Recombination within the myelin basic protein gene created the dysmyelinating shiverer mouse mutation. Proc. Natl. Acad. Sci. U.S.A. 83:7542-7546

Musse A, Gao W, Homchaudhuri L, Boggs J, Harauz G (2008) Myelin Basic Protein as a "PI(4,5)P2-Modulin": A New Biological Function for a Major Central Nervous System Protein. Biochemistry, ahead of print

Nabet A, Boggs JM, Pézolet M (1994) Study by infrared spectroscopy of the interaction of bovine myelin basic protein with phosphatidic acid. Biochemistry 33:14792-14799

Olivotto M, Arcangeli A, Carlá M, Wanke E (1996) Electric fields at the plasma membrane level: a neglected element in the mechanisms of cell signalling. Bioessays 18:495-504

Omlin FX, Webster HD, Palkovits CG, Cohen SR (1982) Immunocytochemical localization of basic protein in major dense line regions of central and peripheral myelin. J. Cell Biol. 95:242-248

Ono A, Ablan SD, Lockett SJ, Nagashima K, Freed EO (2004) Phosphatidylinositol (4,5) bisphosphate regulates HIV-1 Gag targeting to the plasma membrane. Proc. Natl. Acad. Sci. U.S.A.101:14889-14894

Padmore L, Radda GK, Knox KA (1996) Wortmannin-mediated inhibition of phosphatidylinositol 3-kinase activity triggers apoptosis in normal and neoplastic B lymphocytes which are in cell cycle. Int. Immunol. 8:585-594

Pedraza L, Fidler L, Staugaitis SM, Colman DR (1997) The active transport of myelin basic protein into the nucleus suggests a regulatory role in myelination. Neuron 18:579-589

Pfeiffer SE, Warrington AE, Bansal R (1993) The oligodendrocyte and its many cellular processes. Trends Cell. Biol. 3:191-197

Pike LJ (2004) Lipid rafts: heterogeneity on the high seas. Biochem. J. 378:281-292

Polito A, Reynolds R (2005) NG2-expressing cells as oligodendrocyte progenitors in the normal and demyelinated adult central nervous system. J. Anat. 207:707-716

Powis G, Bonjouklian R, Berggren MM, Gallegos A, Abraham R, Ashendel C, Zalkow L, Matter WF, Dodge J, Grindey G (1994) Wortmannin, a potent and selective inhibitor of phosphatidylinositol-3-kinase. Cancer Res. 54:2419-2423

Prescott ED, Julius D (2003) A modular PIP2 binding site as a determinant of capsaicin receptor sensitivity. Science 300:1284-1288

Privat A, Jacque C, Bourre JM, Dupouey P, Baumann N (1979) Absence of the major dense line in myelin of the mutant mouse "shiverer". Neurosci. Lett. 12:107-112

Rapedius M, Fowler PW, Shang L, Sansom MSP, Tucker SJ, Baukrowitz T (2007) H bonding at the helix-bundle crossing controls gating in Kir potassium channels. Neuron 55:602-614

References

Raucher D, Stauffer T, Chen W, Shen K, Guo S, York JD, Sheetz MP, Meyer T (2000) Phosphatidylinositol 4,5-bisphosphate functions as a second messenger that regulates cytoskeleton-plasma membrane adhesion. Cell 100:221-228

Readhead C, Popko B, Takahashi N, Shine HD, Saavedra RA, Sidman RL, Hood L (1987) Expression of a myelin basic protein gene in transgenic shiverer mice: correction of the dysmyelinating phenotype. Cell 48:703-712

Riccio P, Fasano A, Borenshtein N, Bleve-Zacheo T, Kirschner DA (2000) Multilamellar packing of myelin modeled by lipid-bound MBP. J. Neurosci. Res. 59:513-521

Richter-Landsberg C, Heinrich M (1996) OLN-93: A new permanent oligodendroglia cell line derived from primary rat brain glial cultures. J.Neurosci. Res. 45:161-173

Rispoli P, Carzino R, Svaldo-Lanero T, Relini A, Cavalleri O, Fasano A, Liuzzi GM, Carlone G, Riccio P, Gliozzi A, Rolandi R (2007) A Thermodynamic and Structural Study of Myelin Basic Protein in Lipid Membrane Models. Biophys. J. 93:1999-2010

Rivas AA, Castro RM (2002) Interaction of bovine myelin basic protein with triphosphoinositide. J. Colloid Interface Sci. 256:290-6

Ro H, Carson JH (2004) pH microdomains in oligodendrocytes. J. Biol. Chem. 279:37115-37123

Roach A, Boylan K, Horvath S, Prusiner SB, Hood LE (1983) Characterization of cloned cDNA representing rat myelin basic protein: absence of expression in brain of shiverer mutant mice. Cell 34:799-806

Runnels LW, Yue L, Clapham DE (2002) The TRPM7 channel is inactivated by PIP(2) hydrolysis. Nat. Cell Biol. 4:329-336

Rusten TE, Stenmark H (2006) Analyzing phosphoinositides and their interacting proteins. Nat. Methods 3:251-258

Shanshiashvili LV, Suknidze NC, Machaidze GG, Mikeladze DG, Ramsden JJ (2003) Adhesion and clustering of charge isomers of myelin basic protein at model myelin membranes. Arch. Biochem. Biophys. 419:170-177

Sheetz MP, Sable JE, Döbereiner H (2006) Continuous membrane-cytoskeleton adhesion requires continuous accommodation to lipid and cytoskeleton dynamics. Annu. Rev. Biophys. Biomol. Struct. 35:417-434

Shi S, Jan LY, Jan Y (2003) Hippocampal neuronal polarity specified by spatially localized mPar3/mPar6 and PI 3-kinase activity. Cell 112:63-75

Simons M, Krämer EM, Thiele C, Stoffel W, Trotter J (2000) Assembly of myelin by association of proteolipid protein with cholesterol- and galactosylceramide-rich membrane domains. J. Cell Biol. 151:143-154

Simons M, Trotter J (2007) Wrapping it up: the cell biology of myelination. Curr. Opin. Neurobiol. 17:533-540

Smith (1992) The basic protein of CNS myelin: its structure and ligand binding. J. Neurochem. 59:1589-1608

References

Smith R (1977) Non-covalent cross-linking of lipid bilayers by myelin basic protein: a possible role in myelin formation. Biochim. Biophys. Acta 470:170-184

Sogabe K, Roeser NF, Davis JA, Nurko S, Venkatachalam MA, Weinberg JM (1996) Calcium dependence of integrity of the actin cytoskeleton of proximal tubule cell microvilli. Am. J. Physiol. 271:F292-303

Stevens B, Porta S, Haak LL, Gallo V, Fields RD (2002) Adenosine: a neuron-glial transmitter promoting myelination in the CNS in response to action potentials. Neuron 36:855-868

Taveggia C, Thaker P, Petrylak A, Caporaso GL, Toews A, Falls DL, Einheber S, Salzer JL (2008) Type III neuregulin-1 promotes oligodendrocyte myelination. Glia 56:284-293

Tekkök SB, Faddis BT, Goldberg MP (2005) AMPA/kainate receptors mediate axonal morphological disruption in hypoxic white matter. Neurosci. Lett. 382:275-279

Tekkök SB, Ye Z, Ransom BR (2007) Excitotoxic mechanisms of ischemic injury in myelinated white matter. J. Cereb. Blood Flow Metab. 27:1540-1552

Trajkovic K, Dhaunchak AS, Goncalves JT, Wenzel D, Schneider A, Bunt G, Nave K, Simons M (2006) Neuron to glia signaling triggers myelin membrane exocytosis from endosomal storage sites. J. Cell Biol. 172:937-948

Trapp BD, Kidd GJ (2004) Structure of myelinated axons. In: Myelin biology and disorders (Lazzarini RA), pp 3-22. San Diego: Elsevier Academic Press

Uversky VN (2002) What does it mean to be natively unfolded? Eur. J. Biochem. 269:2-12

Várnai P, Balla T (1998) Visualization of phosphoinositides that bind pleckstrin homology domains: calcium- and agonist-induced dynamic changes and relationship to myo-[3H]inositol-labeled phosphoinositide pools. J. Cell Biol. 143:501-510

Várnai P, Balla T (2007) Visualization and manipulation of phosphoinositide dynamics in live cells using engineered protein domains. Pflugers Arch. 455:69-82

Vemuri GS, McMorris FA (1996) Oligodendrocytes and their precursors require phosphatidylinositol 3-kinase signaling for survival. Development 122:2529-2537

van der Wal J, Habets R, Várnai P, Balla T, Jalink K (2001) Monitoring agonist-induced phospholipase C activation in live cells by fluorescence resonance energy transfer. J. Biol. Chem 276:15337-15344

Wang J, Gambhir A, McLaughlin S, Murray D (2004) A computational model for the electrostatic sequestration of PI(4,5)P2 by membrane-adsorbed basic peptides. Biophys. J. 86:1969-1986

Waxman SG (1997) Axon-glia interactions: building a smart nerve fiber. Curr. Biol. 7:R406-10

White R, Gonsior C, Krämer-Albers E, Stöhr N, Hüttelmaier S, Trotter J (2008) Activation of oligodendroglial Fyn kinase enhances translation of mRNAs transported in hnRNP A2-dependent RNA granules. J. Cell Biol. 181:579-586

Yang JC, Chang PC, Fujitaki JM, Chiu KC, Smith RA (1986) Colvalent linkage of phospholipid to myelin basic protein: identification of phosphatidylinositol bisphosphate as the attached phospholipid. Biochemistry 25:2677-2681

Yeung T, Gilbert GE, Shi J, Silvius J, Kapus A, Grinstein S (2008) Membrane Phosphatidylserine Regulates Surface Charge and Protein Localization. Science 319:210-213

Yeung T, Ozdamar B, Paroutis P, Grinstein S (2006a) Lipid metabolism and dynamics during phagocytosis. Curr. Opin. Cell Biol. 18:429-437

Yeung T, Terebiznik M, Yu L, Silvius J, Abidi WM, Philips M, Levine T, Kapus A, Grinstein S (2006b) Receptor Activation Alters Inner Surface Potential During Phagocytosis. Science 313:347-351

Zeller NK, Hunkeler MJ, Campagnoni AT, Sprague J, Lazzarini RA (1984) Characterization of mouse myelin basic protein messenger RNAs with a myelin basic protein cDNA clone. Proc. Natl. Acad. Sci. U. S. A. 81:18-22

Zhang S (2001) Defining glial cells during CNS development. Nat Rev Neurosci 2:840-843

Zimmerberg J, McLaughlin S (2004) Membrane curvature: how BAR domains bend bilayers. Curr. Biol 14:R250-2

I want morebooks!

Buy your books fast and straightforward online - at one of world's fastest growing online book stores! Environmentally sound due to Print-on-Demand technologies.

Buy your books online at
www.morebooks.shop

Kaufen Sie Ihre Bücher schnell und unkompliziert online – auf einer der am schnellsten wachsenden Buchhandelsplattformen weltweit! Dank Print-On-Demand umwelt- und ressourcenschonend produziert.

Bücher schneller online kaufen
www.morebooks.shop

KS OmniScriptum Publishing
Brivibas gatve 197
LV-1039 Riga, Latvia
Telefax: +371 686 204 55

info@omniscriptum.com
www.omniscriptum.com

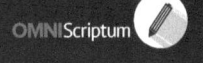

Printed by Books on Demand GmbH, Norderstedt / Germany